Shaping Ecology

Shaping Ecology: The Life of Arthur Tansley

Peter Ayres

WILEY-BLACKWELL
A John Wiley & Sons, Ltd., Publication

British Ecological Society

New Phytologist

This edition first published 2012 © 2012 by John Wiley & Sons, Ltd

Wiley-Blackwell is an imprint of John Wiley & Sons, formed by the merger of Wiley's global Scientific, Technical and Medical business with Blackwell Publishing.

Registered Office
John Wiley & Sons, Ltd, The Atrium, Southern Gate, Chichester, West Sussex, PO19 8SQ, UK

Editorial Offices
9600 Garsington Road, Oxford, OX4 2DQ, UK
The Atrium, Southern Gate, Chichester, West Sussex, PO19 8SQ, UK
111 River Street, Hoboken, NJ 07030-5774, USA

For details of our global editorial offices, for customer services and for information about how to apply for permission to reuse the copyright material in this book please see our website at www.wiley.com/wiley-blackwell.

Library of Congress Cataloging-in-Publication Data

Ayres, P. G. (Peter G.)
Shaping ecology : the life of Arthur Tansley / Peter Ayres.
      p.   cm.
   Includes bibliographical references and index.
   ISBN 978-0-470-67156-6 (hardcover) – ISBN 978-0-470-67154-2 (pbk.)   1. Tansley, A. G. (Arthur George), Sir, 1871–1955.   2. Tansley, A. G. (Arthur George), Sir, 1871–1955–Political and social views.   3. Ecologists–Great Britain–Biography.   4. Botanists–Great Britain–Biography. 5. Ecology–Great Britain–History.   6. Nature conservation–Great Britain–History.   7. Natural areas–Great Britain–History.   8. Nature Conservancy (Great Britain)–History.   9. Field Studies Council (Great Britain)–History.   10. Great Britain–Environmental conditions.   I. Title.
   QH31.T37A87 2012
   577.09–dc23
                                            2011040963
A catalogue record for this book is available from the British Library.

Wiley also publishes its books in a variety of electronic formats. Some content that appears in print may not be available in electronic books.

Set in 10/12.5pt Minion by SPi Publisher Services, Pondicherry, India
Printed and bound in Malaysia by Vivar Printing Sdn Bhd

1   2012

Contents

Figures

More pictures of Tansley may be seen at www.newphytologist.org/tansley/, and
of his family at www.branscombeproject.org.uk/Attic%20Trunks.pdf.

Foreword

by Alastair Fitter FRS on behalf of the New Phytologist Trust and the British Ecological Society

Arthur Tansley was one of the few people who can claim the distinction of being one of the founders of a discipline. He was present at the birth of the modern discipline of ecology, a science whose name was only coined – in German – at the end of the 19th century. Tansley was a man of vision but also a great moderniser, whose influence is still strongly felt. He was a botanist and something of an iconoclast, who railed against the rigid and old-fashioned teaching methods of his day. His solution was also very modern: he harnessed media power by founding and editing a new journal, albeit one with a very traditional name – the *New Phytologist* – which is today one of the most successful international plant science journals. Not content with that, he was instrumental in converting a committee set up to study British vegetation into the world's first learned society for ecologists, the British Ecological Society (BES), and became its first President (and also the only person yet to have been President twice!) and more importantly the editor of its journal, the *Journal of Ecology*, which similarly has survived to be among the leading journals today. That trio of achievements gives him an unique legacy in British science, since all have prospered. At the time, too, they gave him a remarkable and very beneficial influence on the way that ecology developed in the UK.

Tansley had a genuinely international outlook and was in close touch with botanists and ecologists (not that they called themselves that at first) in the rest of Europe and in the USA. He was also a man of diverse interests and in the period after World War I became intensely interested in psychoanalysis. He held teaching posts at both Oxford and Cambridge and so influenced generations of students, and during and after World War II played a major role in the development of the modern conservation movement in Britain, jointly chairing (with Julian Huxley) the Wildlife Conservation Special Committee, whose report led to the establishment of national parks in England and to the creation of the Nature Conservancy and the beginnings of a national approach to nature conservation. My own link with Tansley, apart from having been an editor of both of 'his' journals and President of the BES, is that my father was secretary to the Committee.

This period was a productive one for Tansley, freed from some of his earlier commitments. In 1939 he also produced his magnum opus *The British Islands and their Vegetation*, a tour de force that distilled 50 years' experience of studying the landscape and shaped the thinking of British plant ecologists for 50 years; it is strikingly modern in its understanding of the role of environment in ecology. His far-sightedness is perhaps best illustrated by his coining of the word 'ecosystem' in 1935, an insight that enabled the discipline of ecosystem science to develop, mainly in the USA, and which underpins the modern concept of ecosystem services – the goods that we as humans receive for free from the natural world.

Science tends to have a short memory. Research papers rarely cite work that is more than 20 years old and Tansley's work is no exception to that. But the lack of direct mention conceals an enormous debt that the modern discipline owes to him and his is one of the few names from that generation familiar to the current one. The New Phytologist Trust (which owns the journal) has already celebrated its centenary and created an outstanding series of Tansley review articles to honour its founder, and the BES will follow suit in 2013, hosting the International Congress of Ecology to mark its 100th annual meeting; again, Tansley's name will be much in evidence. But for a wider audience, Peter Ayres has done a great service in providing this account of a great man and how he came to achieve so much, in a manner which should make his achievements accessible to anyone who wants to understand the intellectual roots of modern ecology and conservation science.

Alastair Fitter

Preface and Acknowledgements

As the young science that was to be called ecology emerged in the first half of the 20th century its development was indelibly shaped by Arthur Tansley. His name is, however, largely unrecognised and few beyond a relatively small group of professional ecologists are aware of the priceless practical legacy that he left to everyone who cares about wildlife and its conservation. In writing this book I have tried to show why all of us – not just Britons – should be grateful to him and remember his name.

As I researched Tansley's life I discovered a man whose influence extended beyond ecology, to psychology and philosophy. I discovered too a man who fought effectively to defend freedom in science when it was under threat. The major events in his life are described in several obituaries written by his friend, Harry Godwin, but these only whet the appetite for they give little impression of the Arthur Tansley behind the published works. Moreover, they largely ignore the context of his life – the times in which he lived. Times that included two world wars and several economic depressions, events that critically affected political and public attitudes towards what he was trying to achieve. Encouraged by those who knew far more about Tansley's contribution to ecology than I did, I embarked on the task not just of making his achievements better known but of presenting a broader picture of this complex man and his world.

I had learned a little about the man after being asked to help his edit 'his' journal, the *New Phytologist*, and to join its Trust. A fellow trustee and ex-student of Tansley's, Jack Harley, would sometimes recount his personal experiences of the great man – always amusingly, and usually after a dinner and wines of which Tansley would have approved. It was, however, only after I met Tansley's grandchildren, Margaret Lythgoe-Goldstein and Martin Tomlinson, that I was able to grasp a fuller picture of the man. They shared with me memories of their grandfather – not always happy – correcting any false impressions that I had, and they helped me access family letters and photographs. I thank them and other members of the family, Alice Lythgoe-Goldstein, Louisa Tomlinson, and Peter Dickens, for their help. Where family matters are concerned, I thank also

Branscombe's (Devon) historians, Barbara Bender and John Torrance, for guiding me through hundreds of letters remaining at an old family home in that village. I thank Jennifer Newton (née Clapham) for allowing me to quote from her father's letters. Finally, I thank Donald Pigott – probably the last ecologist alive to have worked with Tansley – for completing my picture of Tansley.

Laura Cameron (Queen's University, Ontario) and John Forrester (University of Cambridge) gave me invaluable help where Tansley's involvement with psychology was concerned. Andrew Smith, Robin Darwal-Smith, and Stephen Harris (University of Oxford) escorted me through Tansley's Oxford years, while Simon Bailey, Keeper of the Archives at the Bodleian library was unfailingly helpful. I owe special thanks for individual pieces of research to Christine Alexander and David Briggs (Plant Life Sciences, Cambridge), Chris Jakes (Cambridgeshire County Library), and Michelle Losse (archivist, Royal Botanic Gardens, Kew). Enthusiastic help was received also from Mark Seward (University of Bradford), Rebecca Farley (Field Studies Council), Andrew Roberts (University of Middlesex), and David Elliston Allen. I thank David Wilkinson (Natural England) for his guidance concerning the recent history of nature reserves and protected sites. Several of those named above, plus Malcolm Latham and Rich Norby, helped further by commenting on draft chapters.

John Sheail was generous with ideas, advice, and information about Tansley and the people who surrounded him. Above all, I thank John for the encouragement he gave me to complete the project, and my wife, Mary, for her unfailing support throughout its considerable length.

I am most grateful to the New Phytologist Trust and the British Ecological Society for their financial support.

I hope I have not distorted the truth for, as Tansley wrote,

> We must never conceal from ourselves that our concepts are the creation of the human mind which we impose on the facts of nature.
>
> *Tansley 1920, p.120.*

Arthur Tansley photographed during the 1930s when he was Sherardian Professor of Botany at the University of Oxford. (By permission of the Cambridgeshire Collection, Cambridge Central Library.)

1 Kingley Vale: Worth Fighting For

While fleeing with the remnants of his army from defeat at the Battle of Worcester in 1651, the future King Charles II rode along what is today a long distance footpath, appropriately named Monarch's Way. Local tradition holds that he reined in his horse above Kingley Vale in West Sussex and, looking down on its beauty, exclaimed 'England is surely worth fighting for'. A similar thought must have crossed the mind of Arthur Tansley many times, for Kingley Vale was an inspiration through his long life, a place to which at critical moments he returned in spirit, and often in person.

Tansley regarded the view from Kingley Vale as 'the finest in England'.[1] Situated towards the western end of what is now the South Downs National Park, its natural beauty is imbued with a strong sense of history. The Vale's wooded entrance leads to a huge green amphitheatre; a steep-sided horseshoe whose open end faces southward. At its heart are ancient yews, some more than 1000 years old (Figure 1.1). On its mottled green flanks, younger yews mix with ash. Hawthorn and scattered juniper bushes colonise the higher slopes. The celebrated view is best enjoyed from its closed and higher end, where rabbits and deer keep the chalk-loving grasses and herbs free from invading scrub. More than 50 kilometres of Britain's south coast is displayed; to the southeast lies the old Roman city of Chichester, the spire of its 12th century cathedral rising 85 metres above the coastal plain; to the south-west lies the entrance to Portsmouth harbour, home of the Royal Navy and of Nelson's flagship, *Victory*. The silence is disturbed only by the songs of birds and the sound of the wind funnelling up the horseshoe.

Tansley had fallen in love with what he liked to call the 'great hills of the South country'[2] when his parents sent him away from home in London to

Shaping Ecology: The Life of Arthur Tansley, First Edition. Peter Ayres.
© 2012 by John Wiley & Sons, Ltd. Published 2012 by John Wiley & Sons, Ltd.

Figure 1.1   Ancient yews at Kingley Vale.

board at Westbury House Preparatory School in Worthing, a small town on the Sussex coast. For the 11-year-old boy, outings to the nearby South Downs – many of whose hills are crowned with Iron Age forts constructed around 300 BC – were an exciting, keenly anticipated relief from the confines of school life. Among these open grasslands, young Arthur could find freedom and indulge his growing fascination with plants, an interest that – thankfully, for the history of ecology – was encouraged by the school's Principal, Mr J. H. Bloom. Arthur shared his enthusiasm for botany with his sister, Maud, six years his elder:

> I have not been able to get any plants yet, but I hope to be able to tomorrow when (if it is fine) we are going up Cissbury [Ring, or fort].
>
> *Arthur to Maud Tansley, 16 April 1883,*
> *from 12 Montague Place, Worthing*[3]

> The plants you sent me were mostly rare and one of them, the yellow dead-nettle, Mr Bloom has not got. The (1) pea and (2) spurge which I send you are (1) a rare one which I should like a better specimen of. And (2) an unnameable spurge on account of the badness of the specimen of which I should like a better specimen. I should like two specimens of the yellow dead-nettle.
>
> *Arthur to Maud Tansley, 6 August 1884, address 'at home'*[4]

Writing to Mr Bloom during the summer holidays of 1884, the boy's tone was extraordinarily mature:

The other day we had a jolly time on the downs not far from Dorking and I turned up a good many plants. Galium cruciatum and G. saxatile, Geum urbanum, Lysimachia nemorum, Vicia sativa var. angustifolia … [the list goes on. He adds] Is Cephalanthera grandiflora rare? They do not say in my floras.

The next time I go to Mill Hill I will be sure and get you a plant of Allium ursinum.

Some of the plants you sent me are labelled Langley Barn. I do not remember it, is it somewhere up Sompting way [near Worthing]? I should think it must be a good place for plants.

Arthur to J. H. Bloom, 2 July 1884, from 167 Adelaide Road, London[5]

In 1884, Arthur contributed to the school's magazine, The Westbury House Ephemeris, an article on the genus *Potentilla* (cinquefoils). The subject had local interest for at least two species, *P. sterilis* (barren strawberry) and *P. reptans* (creeping cinquefoil), could be found at the edge of woodlands scattered among the grassy uplands running east–west, a little north of Worthing. Showing 'astonishing maturity',[6] he compared the taxonomic treatment given to them by four sources, Bentham's *Flora*, the seventh edition of the *London Catalogue*, Hooker's *Student's Flora*, and Babington's *Manual*. Remarkably, Tansley's last research paper, published in 1948 when he was 77 years old, was on 'The nature and range of variation in the floral symmetry of *Potentilla erecta*'.

When aged 25, Arthur Tansley designed a bookplate for his personal library. Seen through an imagined study window, it is a picture of a distant Iron Age hill fort, Chanctonbury Ring, high on the South Downs, 20 miles east of Kingley Vale (Figure 1.2). (The philosopher Bertrand Russell, a friend from Tansley's student days, had once paused on a walk that he and Tansley were taking together to remark, 'I think it may be taken as an axiom, Tansley, that any view that includes Chanctonbury Ring is a good view.'[7]) The bookplate reveals much about Tansley's early life and interests. At its centre is Tansley's desk, upon which there is a contemporary microscope and a magnifying lens, and on a nearby chair a plant-filled vasculum – all tools of his botanist's trade. The upper panels of the window display what he was proud to recognise as critical influences on his education, Highgate School, the London Working Men's College, and Trinity College, Cambridge. In the frame are carefully drawn flowers representative of woodlands, heath, and waysides. There is a selection of books by the authors he revered most, from Shakespeare and Balzac to Matthew Arnold and Shelley, and, from his botanical world, there are books by Charles Darwin and the German masters, Julius von Sachs, Anton de Bary, and Wilhelm Pfeffer. Each detail of the bookplate represents an important facet of the first quarter century of Tansley's life but, most revealingly, at its centre is the British countryside, bringing the whole picture together.

***

Figure 1.2   The bookplate designed by Arthur Tansley. (By permission, Archives of the New Phytologist Trust.)

The loss of Britain's countryside to sprawling cities, and the invasion of its wilder places by the canals and railways connecting those cities, had been lamented by numerous writers as industrialisation gathered pace in the 19th century. Britain may have led the Industrial Revolution but it had not led the world in protecting its own flora and fauna (perhaps because those who could have taken action, the politicians and landowners, had most to lose). In the second half of the 19th century there had been some progress in protecting commons and footpaths, largely within urban areas, and in 1896 the newly formed National Trust began to acquire land, but it was not until 1912 that moves were made *specifically* to establish nature reserves.

On 18 December 1912 *The Times* carried a leading article, much of it drafted by Charles Rothschild,[8] describing an address that Dr Chalmers Mitchell had recently given to the Zoological Section of the British Association at Dundee. He had invited support for the formation of a Society for the Promotion of Nature Reserves (SPNR), reflecting:

> It is only by the deliberate and conscious interference by man … that the evil wrought by man has been arrested … Each generation is the guardian of the resources of the world; it has come into a great inheritance but only as a trustee.

Mitchell had pointed out that while 'primitive races' of man were protected in the United States, northern Scandinavia, and parts of the British Dominions, and, in Germany in particular, land had been set aside for wild-life, there were no such refuges, either for man or for the protection of wildlife in Britain.

The SPNR was a product of the ambition, drive, and financial support of Rothschild, a distinguished entomologist and member of a wealthy family of bankers. As a keen supporter of the National Trust, he had helped it acquire for the cost of £10 the first of its several slices of land at Wicken Fen in Cambridgeshire (1899). He was frustrated, however, feeling that the Trust had no clear policy for the selection and acquisition of nature reserves, and that conservation was not high enough on its list of concerns. He believed the SPNR should work in conjunction with the National Trust, identifying sites which the Trust should acquire for reserves. His vision was that once a reserve had been acquired it should be managed by knowledgeable wardens.[9] In practice, how-ever, he found that after acquiring 300 acres of Woodwalton Fen in 1910, the Trust, with recent experience of the high cost of managing Wicken Fen, was unwilling to accept the additional land. So, he gave Woodwalton to the SPNR together with a bequest of £5000 towards its running costs, although even this sum proved barely sufficient and seriously handicapped the SPNR's capacity to take on further reserves.[10]

Membership of the SPNR was by invitation and was an honour. The first List of Members, published in 1914, included Tansley as one of seven distinguished academic botanists who were to be members of its Council, the balance of botanists and zoologists on the Council not reflecting Mitchell's zoologically orientated rhetoric.[11]

Tansley was an obvious choice because he was President of the British Ecological Society (BES), which only one year earlier had grown out of the British Vegetation Committee, a body whose original aims were to survey the main vegetation types of the British Isles. It was a continuation of such survey work, and its extension throughout the British Empire, that Rothschild saw as a necessary precursor to the selection of reserves. Although lists of potential reserves were drawn up, Britain and its Empire were soon embroiled in World War I (WWI), limiting the fulfilment of the SPNR's plans. The Society survived Rothschild's death in 1923 (and still survives today as the Royal Society of Wildlife Trusts), but it never assumed a leading role in the conservation movement. Also, like other bodies similarly attempting to protect natural resources in the years of economic depression that followed WWI, the SPNR struggled because neither the mood of the public nor the attention of politicians was yet ready to engage with their concerns.[12,13]

The patriotism that always accompanies war resurfaced in the early 1940s as broadcasters, film makers, and journalists all echoed the cry around which Frank Newbould based his celebrated bucolic posters, commissioned by the War Office, 'Your Britain: Fight for it Now!'. In step with the mood of the time, publishers such as Collins commissioned cheap, popular books with titles that included *The Englishman's Country* and *Wild Nature in Britain*, their implicit message being that Britons were fighting not just for their freedom and traditions but for a land whose countryside was extraordinarily beautiful. Growing awareness of the richness and diversity of the British countryside brought with it the perception that post-war industrial recovery posed a new threat, a fear fuelled by the idea that somewhere in the past there had been a Golden Age in the British countryside.

> ... the countryside of Britain reached its supreme beauty at the beginning of the nineteenth century. The fine trees, planted in such numbers fifty or a hundred years earlier, had reached maturity; squalid industrial towns had hardly begun to encroach upon the country ... Since then there has been a general decline.
>
> J. Pennington, *The British Heritage*[14]

> The industrial revolution and the creation of parks around country houses have taken us down to the later years of the nineteenth century. Since that time, and especially since the year 1914, every single change in the English landscape has either uglified or destroyed its meaning, or both.
>
> W. G. Hoskins, *The Making of the English Landscape*[15]

Even Tansley occasionally succumbed to the use of emotive language to support his arguments, as when he wrote to *The Times* about the conflicting demands of post-war military training and nature conservation:

> It is obviously futile to preserve rural beauty if the country is to succumb to an invader, but it is disastrous if preparations for its defence are to destroy in detail a very large part of its character and charm.
>
> A. G. Tansley, *The Times*, 3 December 1946

Encouraged to fight for their homeland, and to defend its natural beauty, it is not surprising that the British people expected to be more involved in decisions affecting the future of its countryside. Through the 1940s public opinion shifted so that it was at last in sympathy with the aims of those who sought to establish national parks. The politicians were, as usual, sensitive to public opinion but, critically, they were also receptive to the urgings of specialist bodies – such as the SPNR and BES – who sought to establish nature reserves, whether as part of, or separate from, national parks.

Tansley was ideally suited to take a leading role in the fight for nature reserves and nature conservation, not just because of the wealth of his knowledge about the vegetation of the British islands, or the high regard in which he was held by peers, but because he was a realist.

In the mid-1900s, while an assistant to Professor F. W. (Frank) Oliver (Figure 1.3) at University College, London, Tansley had been introduced to the vegetation of the rugged coasts of Norfolk and Brittany. He had seen the rapid changes that resulted as shingle banks were moved under the regular pounding of tides and storms; he saw how plants stabilised the sand banks and invaded the salt marshes that built up as silt was deposited in river estuaries. Change was slower in other landscapes but, from the evidence of his own eyes and what he had learned from others, such as Henry Chandler Cowles and Frederic Clements in North America, it was apparent to Tansley that none was static. Preservation, which is merely the maintenance of the status quo, was not an option. If the beauty of the landscape was to be protected it must be through conservation. Implicit in this positive philosophy was intervention by man and the active management of wildlife populations. For Tansley, ecology should inform conservation by first identifying appropriate objectives, and then the most effective methods for attaining them.

Tansley was a realist in another respect. He realised Britain's 'heritage of wild nature' was largely the outcome of past land use. He recognised the essential roles that farming and forestry had played, and must continue to play, in the evolving countryside. He may have regretted the loss of natural vegetation and complained bitterly that the Forestry Commission 'is destroying the beauty of the still wooded Highland glens by substituting close plantations of exotic conifers for the natural oak, birch and pine'[16] but, ever the pragmatist, he did not shy away from Britain's need for more food and timber in its recovery from WWII. For Tansley, fields and forests were integral parts of the whole countryside and their peculiar ecology deserved its own study. In *The British Islands and their Vegetation* (1939a) he had argued, 'Some form of national planning of a systematic "lay-out" of the whole country, in which various interests are duly considered and adjusted – the rural as well as the urban, the spiritual and aesthetic, as well as the industrial and commercial – is now indeed very urgent'.[17]

Figure 1.3   Harry Godwin, top left, and F. F. 'Fritz' Blackman, top right (by permission of the Royal Society of London); Frank Oliver, bottom left (by permission of University College, London), and Roy Clapham, bottom right (by kind permission of his daughter, Jennifer Newton).

By 1945 change was unavoidable,

> Planning for the preservation of rural beauty must be directed to the deliberate conservation of much of our native vegetation, since this is an essential element of natural beauty ... such planning must be balanced and harmonised with land utilisation for agriculture and forestry ...
>   ... to conserve our native vegetation intelligently and effectively we must understand its nature and behaviour under different conditions, an understanding that is gained through the modern science of plant ecology.
>   A. G. Tansley, *Our Heritage of Wild Nature*[18]

This realism that pervaded Tansley's character was the key to his success in many spheres, not least the political where it put him among that small group of experts who could communicate with notoriously science-shy politicians. He could see the politician's point of view; he could understand their wider responsibilities. In 1945 he was appointed to the government's Wild Life Conservation Special Committee, which was to make the case for National Nature Reserves. He soon took over the chairmanship of that Committee, whose report persuaded the government to set up Reserves that were separate and distinct from National Parks, exactly as Tansley had always wanted.

The founding of the Nature Conservancy in 1952, of which he was the first Chairman, and the Conservancy's establishment of National Nature Reserves (NNRs) (Box 1.1), were the most tangible achievements of Tansley's life. Kingley Vale was included in the Conservancy's first list of Reserves. Acquired henceforward for the nation, Kingley Vale would, like the other Reserves, still be accessible to the public, but would be protected. It would be a site where research and experiment would guide conservation. In 1957 a memorial stone was erected at almost the exact spot in the Reserve from which Tansley had so often enjoyed the view. The memorial is an ancient Sarsen stone and on it a bronze plaque reads: 'In the midst of this nature reserve which he brought into being this stone calls to memory Sir Arthur George Tansley, F.R.S., who during a long lifetime strove with success to widen the knowledge, to deepen the love, and to safeguard the heritage of nature in the British Isles'. The words were written by Max Nicholson, his friend and colleague in the struggle for nature reserves.[19] As they emphasise, Tansley fought for the countryside of the whole of the British islands, not just one small corner of England. And his work shaped the development of ecological science across the world.

***

Box 1.1 Nature conservation areas

Nature conservation is, as appropriate, the responsibility of Natural England (NE), Scottish Natural Heritage (SNH) or the Countryside Council for Wales (CCW). Reserves in Northern Ireland are designated and managed by the Northern Ireland Environment Agency.

National Nature Reserves (NNRs)
Over 350 sites whose wildlife and/or geology are of national importance have been given the status NNR. Wildlife and scientific research come first, but most have some provision for public access. They are either owned or controlled by NE, SNH, or CCW, or are held by approved bodies such as the National Trust or Wildlife Trusts. All are SSSIs (see below).

Local Nature Reserves
Established in law by the National Parks and Access to the Countryside Act of 1949 (not applicable in Scotland), these are sites of local importance. They are owned or controlled by local authorities. Public access is allowed to these 1200+ sites. Some are also SSSIs (see below)

Sites of Special Scientific Interest (SSSIs)
The first were identified by the Nature Conservancy in 1949 so that local authorities could give them legal protection. Over 4000 have been designated. They are now the responsibility of NE, SNH, or CCW, who advise landowners on appropriate management.

National Parks
Including the new South Downs National Park, the New Forest National Park, and the Norfolk and Suffolk Broads with equivalent status, the 15 parks account for almost 10% of the area of England, Scotland, and Wales. Each park is administered by its own National Park Authority – armed with appropriate legal powers – whose remit is to conserve and enhance the park's natural beauty, wildlife, and cultural heritage, while promoting opportunities for the understanding and enjoyment of their special qualities. The public does not have access to all land. Special funds are available to help landowners comply with the managing authority's plans. Special planning regulations apply to new buildings, renovations, change of use, etc.

Areas of Outstanding Natural Beauty (National Scenic Areas in Scotland)

Approaching 100 in number, these relatively large areas are jewels in the landscape crown, but unlike National Parks they do not have their own administrative machinery, nor is recreation one of their primary objectives. Special planning regulations apply to new buildings, renovations, change of use, etc.

Critical to the government's acceptance of the Nature Conservancy and NNRs were Tansley's personal authority, acquired through forty years in which he had defined and led the fledgling science of ecology in Britain, and his innate political skills. From small beginnings – research papers in academic journals describing tissues that conduct water in mosses and ferns, published between 1896 and 1904 - his writing grew in scope and ambition as he turned his attention to vegetation and plant ecology. His *The British Islands and their Vegetation*, completed after retirement from the Oxford Chair of Botany in 1939, was a landmark book that distilled for ecologists his lifetime's work. Ten years later, his *Britain's Green Mantle* (1949) helped popularise ecology among a wider audience, helping it awaken to the idea that observing and understanding the natural world offered an enjoyable relief from endless post-war austerity.

Tansley was especially concerned to attract young people to ecology. For most of his working life he was a university teacher. There are conflicting reports about his abilities and popularity as a lecturer but there is no doubting the lead he took in redefining the botanical curriculum to include 'newer' disciplines, such as ecology, not just in his own university but throughout Britain. One of his goals in retirement was to stimulate the flow of ecologically minded students *into* universities. To encourage the teaching of ecology in schools, he wrote with his friend Price Evans a simple guidebook for teachers, *Plant Ecology and the School* (1946). Often overlooked among his achievements is the fact that he helped set up, and was the first President of, the Council for the Promotion of Field Studies, later called the Field Studies Council. This was a body that established residential study centres where young men and women could experience the challenges and rewards of practical field work, carried out at locations specifically chosen for their scientific interest.

Tansley's interests and passions ranged exceptionally broadly, from botany to psychology, from sociology to ethics. His character was equally complex, with its generous share of apparent contradictions. Sociable and outgoing as a young man – with a young man's normal interests in the opposite sex, sport, music, and the theatre – by the second half of his life, that period naturally concentrated upon by his obituarists, he seems to have turned inward. Max Nicholson, whose acquaintance with Tansley did not begin until the 1920s, when Tansley was already middle-aged, found him 'not easy to know'. In spite of their having worked closely together for more than a decade, Nicholson concluded, 'much about him remained a mystery. Tansley had a profound philosophic bent, which set him apart from his fellows, and at times made for loneliness and melancholy'.[20] His grandchildren recall he was very shy with people whom he did not know and he could end a conversation by abruptly leaving the room. At his most relaxed during field work, a devotion to routine nevertheless travelled with him wherever he went, 'nothing was allowed to deflect him from tea round about four o'clock'.[21]

Even his most devoted friends and admirers, such as Harry Godwin (Figure 1.3), were well aware that there was a prickly side to his character. Tansley 'was full of attractive ironic humour and with a very pungent wit' in Godwin's view, but he recorded one incident involving Tansley being as harsh to a colleague, a university treasurer, as he could be to his own family. The treasurer, himself an outspoken Australian,

> for months afterwards spoke with respectful awe of the terms in which he had been addressed when, without prior consultation, he had chosen to alter the mode of paying Tansley's salary to him.[22]

One of Tansley's PhD students in Oxford, Jack Harley, found his professor, 'conversable on any subject, botanical, general or frivolous'[23] but he illustrated the other side of Tansley's character with an anecdote about a student field trip to the nearby Chiltern hills. To the professor's annoyance the party was joined by a number of nuns wearing full habit. Tansley insisted on leading the party through patches of brambles where the nuns' clothes inevitably became snagged.

The same prickliness could be felt by his closest family who often had to step warily around him. His daughters told how during a long summer vacation spent at Blakeney Point, Norfolk, their father slept in a separate tent whenever he was writing. At these times they took great pains not to disturb him, so avoiding the rough edge of his tongue. They joked, 'father is pregnant again'.[24] His granddaughter, Margaret, remembers her elderly grandfather as 'neurotic, a bit scary', though she loved him dearly. She and his grandson, Martin, agree that at his home, Grove Cottage in Grantchester, his routines were never to be interrupted nor his peace invaded, even by the happy cries of children at play.

Disillusionment with the progress of his career in botany and the end of an extramarital affair helped persuade him in mid-life to study psychoanalysis with Sigmund Freud and to contemplate a different career, away from botany. Though self-doubts clearly assailed him during this period, and may less obviously have always been his companions, he was both bold and innovative at crucial moments in his life. He was unafraid to lead, had the gift of persuasion, and above all an extraordinary talent for melding others into novel organisations, synergising their talents. Tansley was not only the founding President of the British Ecological Society, the first body of its kind in the world, but he belonged to that small handful of scientists – mainly botanists and most of them good friends – who in the first two decades of the 20th century defined the new science of ecology.

He remained modest to the end, painfully aware of his own limitations as a botanist. Although in 1904 he had enthusiastically proclaimed 'Ecology may now be considered almost a fashionable study',[25] he was never fully convinced that either his fellow scientists or the public had accepted the 'ecological outlook on biology' which, he thought, was so 'vitally important, not only to pure science but … to human life and activity in the widest sense'.[26]

Notes

1. Hywel-Davies, Thom 1984, p.397.
2. Godwin 1977, p.4.
3. Branscoll (a collection of family letters found at Branscombe, South Devon).
4. Ibid.
5. Ibid.
6. Godwin 1977, p.2.
7. Godwin 1977, p.4.
8. Rothschild, Marren 1997, p.18.
9. Rothschild 1987, p.164.
10. Rothschild, Marren 1997, p.46.
11. Rothschild 1987, p.162.
12. Sheail 1998, p.6.
13. Rothschild, Marren 1997, p.42.
14. Pennington 1948, p.179.
15. Hoskins 1955, p.238 in 1988 edition.
16. Tansley 1939a, p.192.
17. Ibid., p.192.
18. Tansley 1945, preface.
19. Nicholson 1987, p.162.

20. Ibid., p.161.
21. Godwin 1958, p.6.
22. Godwin 1977, p.24.
23. Archives of the New Phytologist Trust.
24. Godwin 1977, p.25.
25. Tansley 1904a, p.191.
26. Tansley 1939b, p.513.

2 The Origins of Ecology

Ecology is the study of plants and animals as they exist in their natural homes, of their "household affairs" and of the communities they form.

A. G. Tansley, *Britain's Green Mantle*[1]

Few would dispute that the term 'ecology' (*öikologie*) was coined by the German zoologist, Ernst Haeckel, in a publication of 1866. It was used with increasing frequency by biologists in the 1880s and 1890s, gradually creeping into everyday English language in the 20th century, though it is alleged the word reached French dictionaries only in 1956.[2] What exactly the word means has proved endlessly debatable, largely because of ecology's mixed parentage, arising as it did out of biology, geology, and geography. Even as *Britain's Green Mantle* was published, with Tansley's homespun definition of ecology (above), the task of defining the subject was becoming more difficult since ecology was increasingly embracing new disciplines such as biochemistry and environmental physics. To the end of his life (in 1955), Tansley regretted that he and his contemporaries had all too often struggled to convince their peers of ecology's own distinctive identity, and to prove that it was something new, separate, and permanent.

The meaning of 'ecology' became even more confused following the environmental crisis of the 1960s, triggered by the publication of Rachel Carson's *Silent Spring*. Tansley's emphasis on 'study', in the definition above, was forgotten as 'ecology's concepts and methods ... were often lost in the extension of the term to incorporate almost any idea, or ideal, concerning the environment taken as meritorious by some group'.[3] 'Ecology' was hijacked by Green politicians, environmental activists, and even those seeking some quasi-spiritual, nature-based explanation of their own place in the world. While welcoming the increased

Shaping Ecology: The Life of Arthur Tansley, First Edition. Peter Ayres.
© 2012 by John Wiley & Sons, Ltd. Published 2012 by John Wiley & Sons, Ltd.

attention their subject was receiving, the community of ecologists found itself, once again, having to defend ecology's scientifically rigorous, distinctive, and independent nature.

***

The roots of scientific ecology are to be found in the 18th and 19th centuries. They are several, the relative importance of each depending on the judgement of the particular commentator. Thus, histories have been coloured by the origins of the writer because the early practices and tenets of ecology were subtly different on different continents. And practising ecologists have sometimes disagreed with historians of science about their subject's origins.

Two conflicting traditions were recognised by Donald Worster.[4] The first, or 'arcadian tradition', is exemplified by the Reverend Gilbert White's letters in which he recorded the lives of the animals (mainly birds) and plants contributing to *The Natural History and Antiquities of Selborne* (1789),[5] his parish in Hampshire. It is humble and romantic in comparison with the second, or 'imperialist tradition', which concerns itself with function and productivity, the establishment of man's dominion over nature. The latter found early expression in the land-grant universities of North America, with their emphasis on applied science, and is exemplified by Charles Bessey's studies, begun at the University of Nebraska in the 1880s, of the ecology of the grasses upon which local prairie agriculture was based.

Undoubtedly, one of the most vigorous roots of modern ecology was plant geography, the study of the distribution of plants (species, genera, or families) together with the causes and implications thereof. An early landmark was Alexander von Humboldt's *Essai sur la Géographie des Plantes* (1805),[6] in which the German explorer recounted his travels through South America. He was the first to try to explain why particular plants occurred in particular places, suggesting that environmental factors such as temperature, rainfall, and the nature of the soil were determining factors. The implication was that a similar mix of plants was likely to occur together wherever the same particular combination of environmental conditions was found. His travels took him to some of the highest mountain ranges in the world and, with remarkable insight, he suggested that altitude and latitude have equivalent effects on plants.

Not everything in von Humboldt's writings was helpful to future generations, however, for he attached much importance to the physiognomy (external appearance) of plants and landscape, listing 19 different plant forms which included cactus, conifer, liane, and grass. Nevertheless, excepting such occasional cul-de-sacs, von Humboldt's work pointed the way forward, away from the purely descriptive approaches of the 18th century. His ideas were developed by, among others, the

Swiss botanist Augustin-Pyramus de Candolle, who introduced an important new factor, competition, to help explain the distribution of species.[7]

Charles Darwin was an avid reader of the works of both von Humboldt and de Candolle.[8] Whether Darwin was effectively an 'ecologist' is arguable, but he undoubtedly facilitated development of the subject. At the very least he inspired his young admirer Haeckel with the thought that the theory of evolution explained 'simply and consistently' and 'mechano-causally' the relations of organisms to their environment.[9] As Haeckel recognised, Darwin's contribution to the origins of ecology was his laying down of a challenge to future generations of ecologists; he challenged them to explain each organism's fitness to survive and reproduce in the environment in which it is found.

A means of meeting that challenge emerged from the German universities in the second half of the 19th century. It became known as the 'New Botany'. Based on the educational principles of observation, measurement, and experiment (rather than learning through received knowledge), it found its apotheosis in the plant physiology taught in the laboratories of Julius von Sachs in Würtzburg and Anton de Bary in Strassburg. Although, to his regret, Tansley never studied in a German university, the lessons of New Botany had spread to Britain by the time he studied botany at Cambridge. Later, in his efforts to change plant geography into ecology, he continually urged upon his contemporaries the need to go beyond the simple descriptive work that formed the basis of their primary surveys. They should reach out to what he called a second level of explanation, which 'unravels the causes of phenomena',[10] and at the core of which are good physiological measurements (Chapter 6).

His beliefs, Tansley freely admitted, owed much to Andreas Schimper and Eugenius Warming, themselves direct beneficiaries of the phytogeography laid out by von Humboldt and Alphonse de Candolle (the son of Augustin-Pyramus). Warming spent several years working in Brazil before settling in his native Denmark. In his greatest work, *Lehrbuch der ökologischen Pflanzengeographie* (1896) (= *Oecology of Plants*, 1909), he distinguished between 'floristic plant-geography', which, he said, is concerned with the construction of lists of plants growing in defined areas, and 'oecological plant-geography'. The latter seeks to find out which species are commonly associated together in similar habitats, and to explain why (a) each species has its own special habit and habitat, (b) species congregate to form definite communities, and (c) communities have a definite physiognomy. This branch of plant geography, he said, sought to determine how different species achieve the same object by the most diverse methods. For example, they can succeed in a dry environment by having leaves that are either covered with a dense layer of hairs, or a thick layer of wax and no hairs, or, alternatively, by reducing the area of their foliage and having succulent stems.

Schimper was an Alsacien who studied with both de Bary and von Sachs before turning from pure physiology to ecology. He too travelled extensively, from North America to Central Africa, and he died in 1901 at the early age of 45, probably as a result of malaria contracted in the tropics. As he travelled, he mapped vegetation, distinguishing plant types according to their latitude – polar, temperate, or tropical – and recognising that vegetation is dynamic, continually changing, a principle that found its way to the heart of Tansley's ecology.

Schimper's greatest work, *Pflanzen-geographie auf Physiologischer Grundlage* (1898) (= *Plant Geography upon a Physiological Basis*, 1903), bears in its outline a great resemblance to Warming's book. Both texts begin with a detailed examination of the factors affecting plant growth – water, heat, light, air, soil, and animals – before turning to 'formations' (plants of similar form, e.g. the large deciduous tree, whose occurrence is governed by either climatic or edaphic factors) and plant communities. Schimper, like Warming, stressed the connection between leaf anatomy and rates of transpiration but, with an understanding resulting from his training as a physiologist, he recognised that, 'It is not the absolute strength of transpiration but its amount relatively [*sic*] to the water supply that leads to protective mechanisms'.[11] Schimper placed greater emphasis than Warming on the soil and recognised an important principle not previously appreciated: plants may be unable to absorb water for several reasons, there may be little in the soil, it may be present but accompanied by an abundance of salts, such as sodium chloride, or the soil may be frozen. Plants from diverse habitats, such as deserts, salt marshes, and tundra, he concluded, have therefore faced a common problem and have often evolved common solutions.

Tansley learned from both Warming and Schimper; it would be a physiological approach that would provide the answers to the greatest ecological problems. It would provide answers to what, he believed, was the central question for ecology, and one he would never stop asking his fellow ecologists, 'What forces drive change in vegetation?'

Tansley's outlook was affected in another way by 19th century continental botany. His most important single contribution to ecological theory would be his launching and promotion of the 'ecosystem', a concept whereby the climate, soils, plants, and animals are all viewed as parts of an integrated system, each in a functional relationship with the others. He was prompted to launch the ecosystem concept in order to counter the idea of the community as a 'super-organism'. This idea had been promulgated by the American ecologist Frederic Clements in the 1920s and then developed further by the South African John Phillips. However, its origins can be traced back to an older 'holistic' tradition in Germany; a tradition imbibed by, among others, von Humboldt and Oscar Drude (a major influence on Clements). It contrasted with the much more reductionist tradition of Darwin, Alphonse de Candolle, and Warming, which placed emphasis on the behaviour of individual species. Tansley distinguished the latter, *autecology*, from *synecology*, the study of vegetation.

***

Tansley recognised man's influence everywhere in the landscape of the British islands, and in his studies of plant succession – the replacement of one type of vegetation by another – he laid great emphasis on 'deflected' succession. Understanding the past contributed to the realism which, as already mentioned in Chapter 1, he brought to national debates about the competing claims of wildlife, forestry, and agriculture.

His pragmatism had particular significance for the practical development of ecology in Britain. He saw clearly that the future of the subject depended upon ecologists, and they had two great needs: some sort of organisation through which they could agree goals and methods, and journals through which they could exchange the results of their researches. He bravely provided both.

Tansley had grown up in a home environment where he had learned that the world can be shaped by able men and women with the courage to take risks and to lead others, as the next chapter will show.

Notes

1. Tansley 1949, p.22.
2. Lévêque 2003, p.2.
3. McIntosh 1985, p.2.
4. Worster D. 1977. *Nature's Economy: the Roots of Nature*. San Francisco: Sierra Club Books.
5. White G. 1789. *The Natural History and Antiquities of Selborne*. London: Cassell & Co.
6. Humboldt FHA von, Bonpland AJA. 1805. *Essai sur la Géographie des Plantes; accompagné d'un tableau physique des regions équinoxiales*. Paris: Levrault, Schoell & Compagnie.
7. Lévêque 2003, p.16.
8. Ayres 2008, p.55.
9. McIntosh 1985, p.36.
10. Tansley 1904a, p.196.
11. Schimper 1898, p.7 of 1903 translation.

3 George Tansley, Christian Socialism, and the Working Men's College

Arthur George Tansley was born on the 15 August 1871, at 33 Regent Square, St Pancras, London, the second child of George and Amelia (née Lawrence) Tansley. Most of the events that shaped his early life occurred within a small area of central London, which encompasses elegant Bloomsbury. It is bounded in the north by the great railway termini of Euston and St Pancras, in the south by the British Museum, and on the west side by Gower Street and University College, London (UCL) (Figure 3.1). Regent Square remains to this day quiet and leafy, although on three sides the elegant Georgian terraces that once enclosed it have been replaced by ugly apartment blocks. By 1871 many of the wider streets in the area had already been softened by the planting of avenues of plane or lime trees, but it would have taken a great deal of imagination to have found the spirit of wild nature either there or in Adelaide Road, two miles to the east, where the family moved in the spring of 1879.

George's business must have been prospering because its workforce had by 1881 expanded to 21 men and a boy, in addition to which his domestic household boasted a resident cook and a housemaid.[1] The situation of the new home was a little more promising for a potential nature lover. Just to the south was Regent's Park with its popular, albeit manicured, Flower Walk and Botanic Garden. Better still, a brisk 20-minute walk to the north was the wilder

Shaping Ecology: The Life of Arthur Tansley, First Edition. Peter Ayres.
© 2012 by John Wiley & Sons, Ltd. Published 2012 by John Wiley & Sons, Ltd.

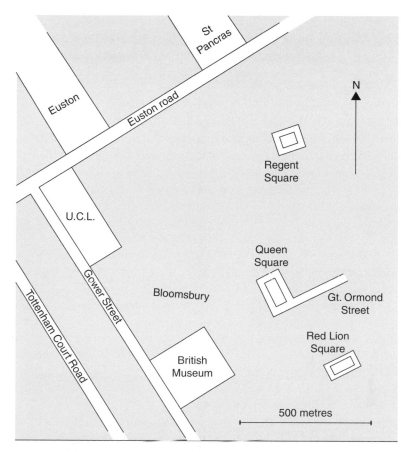

Figure 3.1   Central London, where key events in Tansley's early life occurred.

Hampstead Heath. As the metropolis expanded, the importance of the Heath as 'London's lung' increased. Rising at its highest point over 100 metres above the city centre, it provided city dwellers with fresh air, as well as small hills, streams, wild plants, and open spaces. The delights of the countryside, albeit on a scale that would not have worried a young boy, could be now experienced by Arthur close to his home. A happy and prosperous life beckoned.

To suffer a period of disillusion and disaffection in middle age is not exceptional, and in Tansley's case the causes can be easily identified – frustration both with the progress of his own career and the prospects for ecology, and a crisis in his marriage – but there was ingrained in his character a habit of self-examination and a troubling awareness of his own limitations, either real or imagined. This chapter explores whether the origin of these tendencies lay in his childhood, in particular examining the influence of his parents and of the London Working Men's College (WMC) – founded on Christian Socialist principles – to which they gave so much time and energy.

All the evidence points to Arthur being loved by his parents. In the few biographical notes he left for posterity, his father, George, figures large.[2] George was greatly loved and admired, not only by his son but by his many friends and colleagues at the WMC who saw, 'His purse was as open as his heart … He so perpetually considered his companions' wishes that it was not easy to ascertain his own'.[3,4] Popular and gregarious, George enjoyed nothing better than group activities such as walking, singing, and reading. Involved in cricket and football, George was 'the College's best poetry reader', and was active too in a Shakespeare reading group.[5] Marriage briefly restricted his College activities but they soon picked up again, none more so than his involvement in the College Volunteers (a group rather like the modern Territorial Army Reserve). He helped form a company of the 19th Middlesex Regiment, was an enthusiastic drill teacher, and a superb marksman, steadily progressing through the ranks from private to captain.

The unavoidable inference is that Arthur must often have looked in vain for the company and attention of a father so admirable in everyone's eyes, not least his son's. Although George was passionate about education and, as will shortly be seen, was awarded an honorary degree late in his life, he never studied at a university. He was in awe of the ancient universities of Oxford and Cambridge. Arthur lived his father's dreams when he became a student at Cambridge and later taught at both Cambridge and Oxford.

Amelia was a passionate woman, who doted on Arthur.[6] It is not known whether she was fully in agreement with the decision to send him away to Westbury School, but she certainly showed an anxious regard for his well being. Thus, at the end of his second term, her letters elicited from Mrs Billings (the owner, or matron, of the school) the following reply:

> I will see that he has his milk and half hour walk before breakfast as you wish.
>
> Your dear boy is so clever and industrious, that of course it is a temptation to a master to urge him on – but his health is the first consideration, and, as I know now your wishes in this matter he shall not be pressed.

Mrs Billings made, however, an important exception:

> Arthur will have plenty of opportunity for getting on with Botany in his walks and this will not try him.
>
> *Mrs Billings to Amelia Tansley, 1 May 1884*[7]

Arthur's tendency to shyness and introspection may have had its roots in childhood if, as seems likely, he was too often left to his own devices, or felt he could not live up to his parents' exceptional sociability. Two further factors from his childhood almost certainly encouraged self-contemplation.

First, his sister, Maud, was a problem to her parents; Amelia found her difficult, slovenly, and untruthful. The impression left by numerous family letters is that she was an intense child, possibly unable to cope with any form of stress. She was often sent to live with different families in the country and, much later, after her parents' death, she lived in an institution in Devon. In a time warp, she wore Edwardian clothes until the end of her life. As a child, Arthur could only have been perplexed by his big sister but, as his letters to Maud show, he was fond of her, and treated her with the utmost respect. Thus, in his letter dated 26 August 1897,[8] he discusses topics ranging from sensitive intra-family relationships to William James' recent book on man's need for religious belief[9] – from which it may be inferred she was certainly not simple-minded (the letter is also testimony to the suspicion that in early adulthood Arthur was troubled by the potential contradictions between religious belief and science). As his interest in psychology grew, he must often have pondered the causes of her peculiar nature.

Second, Arthur himself had been born with a disability, a physical one, a deformity of his left hand. Two fingers were fused at the base, their nails badly twisted. He was very conscious of his deformity, hiding the hand in photographs and portraits. Age did nothing to temper his sensitivity for he even snatched it away, rather crossly, when his grandson looked at the hand as it rested on the dining table (M. Tomlinson, personal communication). His student, Jack Harley, believed the hand precluded Tansley from undertaking anything other than simple laboratory experiments, but perhaps this was a blessing in disguise for it was in field botany that he was to excel.

***

When Edith Clements, the wife of a leading American ecologist, met Arthur Tansley many years later, in 1911, she described him as 'a queer mixture of idealist and materialist'.[10] Both elements in his character could be a attributed to a childhood spent in an environment where Christian Socialism was always in the background, so it is time to find out more about that and similar movements.

An essentially idealist movement, Christian Socialism was one of a multitude of socialisms that emerged in the mid-19th century, many of them founded on materialist principles. An ancient philosophy originating with the Greek philosopher, Epicurus, materialism was being re-affirmed both by new discoveries in chemistry, geology, and biology, and by new necessities in politics and economics. Man's view of his relationship with nature was changing.

Materialism asserted that whatever exists is dependent on nature. It was the rejection of idealism, and of a God who periodically intervened to direct nature. In biology, it was the rejection of the 'vital force' many were still calling upon to explain phenomena whose causes could not (at the time) be enumerated. While such materialism would ultimately facilitate Tansley's biology, it happened that when Darwin removed man from the centre of nature, making him merely a part of nature and subject to its laws, he inadvertently let 'social Darwinism' out of Pandora's box. Proposing that the laws of nature were applicable to human society, this philosophy implied that the strongest or fittest should flourish in society, while the weak and unfit should be allowed to founder. (A leading proponent of such elitist theory was Herbert Spencer, whose *biological* works Tansley admired in his youth.[11])

Materialism, together with 19th-century industrialism, helped lay to rest the late 18th-century Romantic Movement inspired by Jean-Jacques Rousseau and his followers. However, while it was no longer possible to sustain the romantic ideal of a countryside inhabited by simple peasants tending a few familial acres and living on the produce of their own labour without the need of external commerce,[12] the Romantics still inspired a fascination with untamed nature that has never diminished. As Bertrand Russell observed in 1946, in the novels of Rousseau's disciples

> we find wild torrents, fearful precipices, pathless forests, thunder-storms, tempests at sea and generally what is useless, destructive, and violent. This change seems to be more or less permanent: almost everybody, nowadays, prefers Niagara and the Grand Canyon to lush meadows and fields of waving corn.
>
> B. Russell, *A History of Western Philosophy*[13]

Nowhere are those storms and torrents seen better than in John Ruskin's romantic sketches of the Scottish Highlands and the English Lake District. Ruskin's wider influence was enormous, not least through William Morris and the attempts of his Arts and Crafts Movement to elevate both the place of nature in design and the social position of the artisan. Most importantly, as will shortly be seen, Ruskin had a strong influence on the lives of two of the founders of the National Trust, Canon Hardwicke Rawnsley and Octavia Hill. At different times, Ruskin, Morris, and Hill all taught at the WMC, countering its materialism with their romantic ideals.

The population of England and Wales more than doubled in the first half of the 19th century and there was a massive net migration from the countryside into the towns and cities. Prior to this, when the bulk of the population had lived in villages and hamlets and worked on the land, there had been little awareness of wild and beautiful places. The land and the plants and animals it supported were merely sources of employment and food. It was when the rural poor left the land to become the urban poor that the distinction between country and city arose. The first moves to protect open spaces within cities and the wider countryside, both for their own sake and for the benefit of the masses, were taken by educated men and women from financially comfortable backgrounds. They were quite naturally opposed by that small group of people who owned most of the land and were fiercely protective of their dominion, resistant to any political interference or suggestion of change. It was the former class of concerned, middle class 'doers', and sometimes even the same individuals, who also sought to improve education for *working* men and women. 'Working' is emphasised because students at the WMC were typically artisans, often self-employed. They had enjoyed little or no formal education but, significantly, they were not so far down the social scale that they lacked the capacity to help themselves. As far as their philanthropic supporters were concerned, these were morally respectable men and women who, if they did not possess it already, could be imbued with the Victorian spirit of self-improvement.

The ethos and day to day operation of the WMC owed much to its first Principal, Frederick Denison Maurice. Maurice was a minister of the Church of England who had been Professor of Theology at King's College, London University, until 1853, when he was dismissed for his dangerous doctrines. His Christian principles had made him a sympathetic supporter of the Chartist movement, which through the 1840s fought for universal male suffrage and the payment of members of parliament (allowing poorer men to stand for election). While its aims were agreed among its leaders, its methods were not. Some advocated violent insurrection, while others, such as Maurice, believed that 'moral force' would win the day. In April 1848, amid widespread unemployment, depressed trade, and working class discontent throughout Europe, the third Chartist Petition was rejected by the House of Commons. Maurice joined with men of similar social conscience, Charles Kingsley (clergyman and novelist) and J. M. F. Ludlow (the intellectual power behind the new movement), in founding the Christian Socialist movement. In the same year they started a night school for working men in Little Ormond Yard, a rough area of Bloomsbury where 'policemen durst not venture alone at night'.[14] The school was a form of social missionary work,[15] designed to teach the 'three Rs' to illiterate students. In 1852 a more ambitious series of lectures and debates was instituted at the Hall of Association in Castle Street East. The last step towards the WMC was precipitated by Maurice's dismissal from King's College. His views had long worried the College's Council but the final straw was his essay, 'On Eternal Life and Eternal Death', published as part of his *Theological Essays* (1853).[16] In the essay he challenged the doctrine of endless punishment for sinners – many of whom, he thought, were poor, helpless people who had suffered enough misery during their earthly lives.

The WMC was not strictly Christian, for its supporters came from all backgrounds, atheist and agnostic included, and it was not strictly socialist for politics played little part in its daily life, but it did embody the ethos of Christian Socialism. The title 'College' symbolised for its members the spirit of a corporate life. As Maurice wrote, it implied 'a Society for fellow work, a Society of which teachers and learners are equal members'. George Tansley became addicted to that spirit, enthusiastically adopting Maurice's dictum, 'A man needs knowledge, not only as a means of livelihood, but as a means of life'.[17]

***

George Tansley's childhood was a sad one. His mother died in 1838, when he was only two years old and his father, Samuel, was in his late forties. Apart from taking long Sunday walks in the countryside with his sons, Samuel was preoccupied with caring for the family business. George and his older brother were sent to a small private day school near their home in Dorset Street, Marylebone, but at the age of 11 he was forced to leave school in order to help in the family business. At about this time his brother died.

The Tansleys were 'ball and rout furnishers'. They provided the furnishings for grand mid-Victorian social occasions, of which there seems to have been no shortage. Their job was to erect marquees, lay down polished dance floors, and provide all the incidental furnishings for the balls and receptions given by London's 'society' hostesses. The business prospered under the stewardship of Samuel and, after his death in 1869, also of George. This was in spite of the latter's fundamental dislike of the extravagance and ostentation of most of his firm's clients. Both Samuel and George were energetic, hard working, and renowned for their eye for detail – all qualities displayed later, in another occupation, by Arthur. The different generations of Tansleys displayed considerable financial acumen. Samuel and George steadily built up the family finances so that when George retired in 1884 the business was sold for about £70 000 (equivalent to more than £5 m today). The wealth accumulated by his immediate forebears gave Arthur that financial independence which, when it suited his purposes, enabled him to live for periods without paid employment while keeping his family in comfort, supported by a small domestic staff.

Arthur Tansley's grandchildren insist that he had 'no interest in politics'. By that they mean he expressed no strong political views, nor was he a member of a political party; which is not to say that he was unaware of political events and opportunities. Similarly, George does not appear to have taken an active part in politics, although he held radical views, once remarking to Arthur, 'If all

England were to vote tomorrow to decide between a monarchy and a republic, I should cast my vote for a republic'.[18]

At the age of 19, in 1855, George was already busy in the family business, but one day he was attracted by an advertisement for the new 'Working Men's College' which had been launched on 30 October of the previous year in Red Lion Square, Bloomsbury. George enrolled himself and in the next three years, during which the WMC moved to larger premises in nearby Great Ormond Street (close to the first Hospital for Children in the English-speaking world, which had opened its doors in 1852), he studied subjects ranging from bible history to maths and mechanics.

George displayed an aptitude for drawing which was soon recognised by his tutor, John Ruskin (Slade Professor of Art in Oxford), who kept one of George's drawings for his own collection.[19] However, it was in mathematics that George excelled, particularly in algebra, and within a year he was helping to teach his fellow students. In 1858 he graduated and became a Fellow of the College, immersing himself in its affairs. In 1863 he married Amelia, whose brother was his good friend, a member of the College, and who probably afforded the pair an introduction. The couple had two children, Maud, born about 1865, and Arthur, born 1871. George was used to balancing the competing demands of the College and his business – which became his sole responsibility when his father, Samuel, died in 1869 – but now he had a growing family to consider. However, far from checking George's involvement with the WMC, it seems that marriage merely encouraged it. George and Amelia's home became renowned in the College's history as a social hub where students and staff could meet.

> All through the winter months we had pleasant opportunities of meeting our College friends at Mr Tansley's house in Regent Square, where Mrs Tansley had an open night once a fortnight for the College circle. There was an informality in these evenings …
>
> *Henrietta Litchfield (née Darwin), 1910*[20]

Both George and Amelia served for ten years on the Council of the Working Women's College, founded in 1864 in Queen Square, adjacent to Great Ormond Street. George taught geometry and algebra at the Women's College while a fellow member of the Council, Octavia Hill, taught reading, writing, and arithmetic, subjects she had already been teaching to women for several years under the umbrella of the Men's College after eagerly joining the WMC in 1856 as secretary to the women's classes.[21] Barely 17 years old, she had been in need of money because her family, who were friends of the Maurices, had fallen on hard times. Most importantly for her, Hill was bewitched by F. D. Maurice's personality and idiosyncratic form of Anglicism.[22]

Practical instruction and lecture courses were given equal emphasis in the WMC. Thanks to Maurice, his friends, and contacts, much of the teaching was given by men who could be regarded as celebrities of their day. Thus, English literature was taught by established authors, such as Thomas Hughes. Classes in drawing were taken by Ruskin, and those in painting by Dante Gabriel Rosetti. William Morris, who had established in the late 1860s a workshop for cabinet making and upholstery in nearby Great Ormond Yard,[23] was finally persuaded in 1880 to give classes in art and design. By 1884, the year he led his little group of socialists 'out into the wilderness' when he formed his own Socialist League,[24] Morris was lecturing on 'socialism' at the WMC.[25] He could have had no shortage of material because in the 1880s socialists were demonstrating a remarkable capacity to disagree with each other and to splinter into factions. This sectarianism was a key reason why the left took so long to become an effective political force.[26]

Morris' brand of socialism was impractical (as, too, was Ruskin's), ultimately being overtaken by other more pragmatic forms. However, as the central figure of the Arts and Crafts Movement, which emphasised the importance of functional design and the use of *nature* as the source for all patterns, whether applied to furniture and fabrics, jewellery, or ceramics, Morris was hugely influential, helping to set the mood and tone of the cultural environment in which Arthur Tansley was growing up.

The WMC had strong links to the old universities, those with Cambridge having been forged as early as 1866 when F. D. Maurice was made that university's Professor of Moral Philosophy. Luckily for the WMC, the chair was not demanding and he was able to continue as Principal of the WMC until his death in 1872. Many of the WMC's tutors had been educated at either Oxford or Cambridge and so, naturally, their experiences of those ancient universities coloured their views of how the WMC should develop. One tutor who had no such experience, but whose reverence for Oxford and Cambridge was unsurpassed, was George Tansley.

> For many years there used to be a College excursion, the favourite places to visit on these excursions being Oxford and Cambridge. Tansley was always keenly interested in these excursions. He loved going to the two old Universities, and it was a great delight to him when his son went in residence at Trinity College, Cambridge, giving him a legitimate excuse for frequent visits to Cambridge.
> These trips were in turn fruitful of good to the Working Men's College.[27]

George proposed that a resident tutor at each of the old universities should be invited to be an *ex officio* member of the Council of the WMC. By the early years of the 20th century, each university was appointing a 'Chief Don' for the

WMC. One of the most successful of these at Cambridge was Francis Cornford (a distinguished classical scholar and husband to the poet, Frances Cornford, Charles Darwin's granddaughter). He not only gave visiting lectures at the WMC but, later, with the help of another young lecturer, G. M. Trevelyan,[28] organised dons in various Cambridge colleges to offer WMC students hospitality during their visits and, even, to arrange football matches and chess tournaments between the WMC and Cambridge students.

Maurice was succeeded as Principal by Thomas Hughes from 1872 until 1881, when in turn Hughes was followed by the exceedingly well connected and influential Sir John Lubbock. Lubbock was the son of a wealthy banker and landowner who was a neighbour of Charles Darwin at Downe in Kent. He was from an early age an enthusiastic botanist and highly talented entomologist. With Darwin's encouragement he pursued his interest in entomology alongside a burgeoning career in banking. Lubbock's researches were rewarded by his election to a Fellowship of the Royal Society. He was not only a celebrity, well known to the readers of *Punch*:

> How doth the Banking Busy Bee
> Improve the shining hours?
> By studying on bank Holidays
> Strange insects and wild flowers[29]

but he was in touch with almost all the leading figures of the day in politics, the City, science, literature, and religion.[30] A member of Thomas Henry Huxley's exclusive 'X Club', Lubbock was later Vice President of the Royal Society, and in 1900 was ennobled as Lord Avebury. Fellow members of the X Club included the Director of the Royal Botanic Gardens at Kew, Joseph Hooker, the celebrated writer on biology and socialism, Herbert Spencer, and the leading physicist, John Tyndall. The latter, together with Huxley himself, was an occasional lecturer at the WMC. Members of the club shared the same philosophy. They were 'united by a devotion to science, pure and free, untrammelled by religious dogma'.[31]

There was another Tansley/Darwin connection involving the WMC. George learned algebra from, and soon became a lifelong friend of, Richard Litchfield, husband of Charles Darwin's daughter, Henrietta. Litchfield and George Tansley were equal in their commitment to, and love of, the WMC. Each found the College early in their life; Litchfield was, at the tender age of 22, among Maurice's first group of ten teachers. A barrister by profession, Litchfield was a gentle, portly man, with a shaggy beard, deeply loved by his Darwin relatives.[32] Like Tansley, he was convinced of the College's role in education outside the classroom. He organised social gatherings for the students and, most notably,

Figure 3.2   George, Arthur (aged 21), and Amelia Tansley at Midsummer Hill, Malvern, 1892. (From the Branscombe Collection, with permission.)

organised a choir and gave singing classes 'for men and girls' on Sundays, often after they had returned from vigorous exercise in the country.

Despite the College and his business making so many demands on George's time, each year in the month of September, when most party givers were away from London, he too would escape the city with his family. A favourite destination was the Malvern Hills (Figure 3.2). The family would walk out together from their cottage high up at West Malvern and George would indulge his other hobbies – sketching and reading.[33] Significantly, Arthur was given by his father a copy of Edwin Lees' *Botany of the Malvern Hills*[34] in August 1884, on the occasion of his 13th birthday. In addition to describing the plants of the Malverns, Lees gave an unusual amount of attention to the geology, physical geography, and climate of the area, emphasising the environmental preferences of individual plant species – an enlightened view of botany that was not lost on his young reader.[35]

It is not certain exactly where and when Arthur's love of botany began – the encouragement of Mr Bloom of Westbury School has already been seen – but, reminiscing when an elderly man, he picked out a member of the WMC, a wood-turner by trade and an excellent and enthusiastic field botanist, who had been especially influential.[36] From the history of the College it seems clear that he was Alfred Grugeon, author of *Botany: Structural and Physiological* (1873) and occasional contributor to the *Geological Magazine*. Yet another person with a lifelong association with the WMC, Grugeon taught there from 1862 to 1869 and also from 1885 until his retirement in 1892, when he became President of the newly formed Lubbock Field Club.[37] A colleague remembered, 'He never used textbooks, and but few diagrams, but in plain words, lit up by a racy humour, explained as no book could the processes of vegetable life, till his pupils thoroughly grasped them. Not only did they understand, they were bitten by his own passion for botany'.[38]

From its earliest days, the WMC had had a tradition of Sunday walks out from London into the countryside, but from 1873 when a Natural History Social and Field Club was formed these became more structured. There were three or four walks every month, many of the botanical ones led by Grugeon (although he was not at the time formally employed by the College). On the walks, he would make collections of plants which he then displayed on return to the College. The spirit of all the walks was meant to be light hearted, giving students a chance to socialise as much as to learn botany or geology. In later years, almost equal numbers of women from the Working Women's College joined the walks and it is highly likely the walking groups were joined by young Arthur Tansley.[39]

After he had sold the family business, and when Ludlow was Principal of the WMC, George Tansley became de facto the chief administrator of the College. George declined any official post or title, despite being pressed by successive principals. It was only in1901, a year before his death, that he finally accepted one, 'Dean of Studies'.[40] As Richard Litchfield noted, 'The well-being of the College was the chief aim, one may say the passion, of his life'.

In January 1889, friends at the College held a party for George and Amelia at which the couple were presented with, among other gifts, a set of 20 volumes of the works of George's old drawing teacher, John Ruskin. On the flyleaf of each volume were printed Litchfield's words, 'A silver wedding gift from colleagues, pupils and friends to George Tansley, M.A., Fellow and Member of Council of the Working Men's College, to commemorate the 3rd of September 1888, being his silver wedding day, and to express the affection and gratitude with which they remember his self-denying and devoted services to the College during 33 years and more'. Significant in itself for what it tells about the love felt by members of the College for the couple, the inscription is significant also for its inclusion of the initials MA. Litchfield had anticipated what was about to happen on 11 May of the same year, the granting to George, by Archbishop Benson, of a Lambeth Degree.[41]

Lambeth Degrees are, to this day, rare and special honours. The Peter's Pence Act of 1533 had given Archbishops of Canterbury the power to grant academic degrees (previously the prerogative of the Pope). It had allowed the Archbishop to override the requirements of the two universities of the day, Oxford and Cambridge, and dispense candidates from residency and, in some cases, examination, in an age when it was difficult to travel to the universities, often because of outbreaks of the plague. This power required confirmation by the Crown, all recipients having to be able to swear an oath to the monarch. George must have overlooked his republican principles; he could have received no higher, or more appropriate, honour.

***

Tansley's parents were thus members of a community of idealists – many, but not all, contributors to the WMC – who shared a belief that their world could be changed for the better. They had success in two ways that were to directly influence his life. They helped promote a rapid expansion of adult education in London that soon spread throughout Britain, and they began the fight to preserve open spaces for the enjoyment of the masses.

Taking adult education first, two new converging social movements were apparent by the mid-point in the 19th century. On the one hand, springing from the new industrial towns, there was an aggressive, class conscious, and coercive labour movement, of which the Chartists were one expression. On the other hand, there was a middle class movement, characterised by its conscience-stricken sympathy for the poor, and speaking of justice and class reconciliation. Adult education provided a convenient strife-free meeting ground for the two. Initially, the middle class organisers sometimes misjudged the needs of the working men and women. Thus, standards at the WMC disappointed Maurice in the early years because his students, over half of whom were manual workers, wanted subjects taught at a more elementary level than he and the other founders had anticipated. An adult school for illiterate students had to be introduced therefore in 1855 and two years later an elementary class was created to form a bridge between the adult school and the College.

George Tansley was always keen to ensure that the courses taken by the brighter students at the WMC would fit them to meet the requirements for entry into the University of London. However, only a handful of students from the WMC, or from comparable institutions, was ever able to take the huge step upward to a full university education. These few exceptions served, of course, to raise the spirits of students and staff alike. None was more notable than Thomas Okey, Cambridge's first Professor of Italian Studies. He was educated at Toynbee Hall,[42] which had been founded by the Reverend Samuel Barnett in

Whitechapel, one of east London's most deprived areas, in 1885. Born to a family of basket-makers, Okey left school at 12, but attended lectures at Toynbee where he learnt Italian, later travelling to that country.[43] The originally limited curriculum of such institutions had clearly diversified, as is also evidenced by the subject of a set of lectures given by Arthur Tansley at Toynbee Hall in 1899; they were on plant geography.[44]

The ideals of the WMC spread quickly from London to the provinces, with a dozen or more similar colleges being founded in England, and two in Scotland, between 1855 and 1868. What happened to the college started in Manchester in 1858 typifies developments in adult education that were occurring in industrialised cities across Britain.

The original Working Men's College in Manchester survived only three years as an independent body before it fused with Owens College, a comparable institution, itself only seven years old.[45] Manchester's population in 1801 was a mere 93 000, but 1871 it had more than quadrupled to 475 000. Out of the industrial grime there grew, however, the phenomenon of 'civic pride'. The wealthy merchant middle classes were fiercely competitive, each wanting for their own city the finest buildings and the noblest institutions. A group of Manchester businessmen raised over £200 000 which, in 1873, allowed the previously cramped Owens College to move into new buildings on a large site at Oxford Road – the hub of the future University of Manchester.

Students taking degree courses in the early years had to register through the University of London, and teaching at Owens had to follow the rules and regulations set down by London. A major step forward was taken in 1880 when the Victoria University was created as a federal institution that joined Owens with similar colleges in Leeds and Liverpool. The city fathers' ambitions were finally realised in 1903 when Manchester University was granted its own separate Royal Charter (Leeds and Liverpool Universities received their charters at the same time).[46]

This growth of colleges and universities, which broadly coincided with an expansion of educational opportunities for women, included the establishment of many courses in biology, each requiring new teaching staff. Among the beneficiaries of such opportunities were men and women who played important parts in the life of Arthur Tansley, as will be seen. Thus, two of his colleagues on the British Vegetation Committee, William Smith and Charles Moss, were connected with Leeds University, the first as a student, the second as a lecturer. Frederick Weiss, a colleague, and Marie Stopes, a student, at University College, London (UCL), became, respectively, Professor and Assistant Lecturer in Botany at Manchester. Ethel Thomas, his research assistant at UCL, held teaching posts at the new University Colleges of Cardiff and Leicester. These were the men and women who, as the momentum of research built up, were to write the original research papers that sustained Tansley's journal, *New Phytologist*, through its first years (chapter 7).

Turning to open spaces and their preservation (conservation came later), an event of great significance had occurred in 1863, eight years before Arthur's birth, when the Commons Preservation Society (CPS) was founded by a young barrister, George Shaw-Lefevre, with the support of T. H. Huxley, John Stuart Mill, Leslie Stephen, and Thomas Hughes. The 'common lands' were areas where for centuries local people had enjoyed the right to graze stock, take wood, pick fruit, or conduct ancient trades such as charcoal burning, but these commons were fast disappearing as they were enclosed – all quite legally. In 1865 the CPS appointed as its Honorary Solicitor, Robert Hunter, a young graduate from UCL. Over the following 20 years, the CPS fought battles to preserve such common lands as Wimbledon Common, Epping Forest, Hampstead Heath, and the Weald (that area lying between the chalk escarpments of the North and South Downs). Most of their battles were won thanks to the brilliant tactics of Hunter and the skilful manoeuvring of the MPs, Shaw-Lefevre and Hughes. Their achievements eventually received approval from the highest level when, in May 1882, Queen Victoria visited Epping Forest to dedicate it 'to the use and enjoyment of the public for all time'.[47]

Apart from her work for the WMC, Octavia Hill had many interests, among them teaching Thomas Hughes' children and copying works of the great artists for John Ruskin. However, her major concern was improving the housing conditions of London's poor, which caused her to look at the wider issues of urban planning. This, in turn, made her aware of a threat to build over Swiss Cottage Fields, where she had played happily as a child. In order to learn how to fight such a threat she joined the CPS and met Hunter. The pair lost this particular battle but Hill was now fully committed to the goals of the CPS and, in time, the war was won; whereas 260 000 acres of common land were enclosed between 1859 and 1869, a mere 26 000 acres were enclosed in the next 20 years.[48]

It was Ruskin who was responsible for introducing Hill to Hardwicke Rawnsley who, along with Robert Hunter, was to be her co-founder of the National Trust. Ruskin had met the Oxford student, Rawnsley, at a College breakfast and immediately drafted him into his team of road-menders – a group of enthusiasts who were repairing Hinksey Road, a muddy lane that linked the city with Hinksey, a village whose inhabitants were mainly poor labourers.[49]

Rawnsley decided to train for the ministry, but struggled unhappily in his first job as an assistant in a hostel for down-and-outs in Soho. With a recommendation from Ruskin, he made himself known to Hill, then at the height of her fame as a housing reformer. For a while, Hill and Rawnsley worked together but the latter soon suffered a nervous breakdown. Hill kindly arranged a period of convalescence for him with her friends who lived in the Lake District. Thus began his love affair with the Lake District, consummated when he was ordained and appointed to the parish of Wray-on-Windermere in 1877.[50]

By the mid-1880s, members of the CPS increasingly recognised that the Society should broaden its remit to include rights of way and, in conjunction with the Kyrle Society,[51] to fight actively for the preservation of beautiful countryside, whether it was common land or in private ownership. Various other bodies, like the Kyrle, were interested in the preservation of ancient buildings and after several years of meetings and discussions between such groups, the National Trust for Places of Historic Interest or Natural Beauty was finally established in 1895. Robert Hunter was the Chair of its Executive Committee. The Reverend Canon Rawnsley was appointed its Honorary Secretary, a post he was to hold for 26 years.

The Trust met those human needs so eloquently described by Octavia Hill in one of her earlier battles to protect the paths and open spaces of London:

> the need of quiet, the need of air, the need of exercise, and ... the sight of the sky and of things growing, seem human needs, common to all men.[52]

Among the Trust's objectives, legally enshrined in a Memorandum of Association drawn up under the Companies Act,[53] was the safeguarding of sites that were important for their wildlife. The first it acquired was a precipitous 4.5 acres of coastline at Dinas Oleu, just behind the small town of Barmouth, in mid-Wales. Rather than being carefully selected, this small, wild, and beautiful space came into the Trust's possession because of personal friendships. It was given to the Trust by Mrs Fanny Talbot, a wealthy philanthropist, who happened to be a friend of Octavia Hill and Canon Rawnsley, as well as of Ruskin. Among the Trust's next acquisitions were (parts of) Wicken Fen in Cambridgeshire and Blakeney Point in Norfolk, both soon to play important parts in Tansley's life. Not all sites were so well chosen, as pointed out by Charles Rothschild (Chapter 1), and their active management was all too often sacrificed in favour of simple preservation. In the acquisition of buildings, the Trust initially confined itself to small, vernacular houses – often recommended to it by William Morris' Society for the Preservation of Ancient Buildings. It was not until the 1930s that its policy changed and larger run-down country houses and castles came to dominate its agenda.[54]

***

In an age when much was talked and written about socialism, Arthur Tansley was able to grow up in an environment where socialism was practised, albeit of a gentle, Fabian kind.[55] The expansion of educational opportunities was generating an awareness among ordinary men and women that they had a right to share in, and even take some responsibility for, the countryside and the natural world. The Working Men's College may have stolen too much of his parents' attention, and encouraged what became a lifetime's habit of introspection, but in other ways it left him a rich legacy that included not merely a love of botany but first-hand evidence that the world could be changed by someone like him.

Notes

1. National Census 1881.
2. Godwin 1957, pp.245–6.
3. C. P. Lucas, p.175 in Davies 1904.
4. Ibid., p.143.
5. F. J. Furnivall, p.55 in Davies 1904.
6. Amelia doted also on the Reverend Alfred Ainger. Numerous notes that she took of his sermons and lectures have been found in the Branscombe Collection.
7. Branscoll (a collection of family letters found at Branscombe, South Devon).
8. Ibid.
9. *The Will to Believe* by William James, a 'most luminous and inspiring writer', according to Tansley (letter dated 26 August 1897, Branscoll). William James was Professor of Psychology at Harvard University and brother of the novelist Henry James. In the same letter to Maud, Tansley asks her to send his 'Sigwort (vol. II)', a book by the German philosopher and logician.
10. Cameron & Matless 2011, p.27.
11. Godwin 1977, p.19.
12. Russell 1946, p.652.
13. Ibid., p.654.
14. J. M. F. Ludlow, p.13 in Davies 1904.
15. Harrison 1961, p.92.
16. Maurice FD. 1853. *Theological Essays*. Cambridge: Macmillan.
17. Harrison 1954, p.21.
18. Ibid., p.120.
19. C. P. Lucas, p.134 in Davies 1904.
20. Litchfield HE. 1910. Richard Buckley Litchfield: a Memoir Written for his Friends by his Wife. Cambridge: privately printed. Available on http://darwin-online.org.uk.
21. C. P. Lucas, p.142 in Davies 1904; Darley 1990, p.68, p.88.
22. Darley 1990, p.44.
23. MacCarthy 1994, p.198.
24. Ibid., p.504.
25. Le Mire 1969, letter of 24 September 1884.
26. Wilson 2002, p.445.
27. C. P. Lucas, p.150 in Davies 1904.
28. Moorman 1980, p.72.
29. Harrison 1954, p.115, citation from *Punch*.

30. Ibid., pp.115–16.
31. Barton 1990, p.57.
32. Spalding 2001, pp.92–4.
33. C. P. Lucas, p.143 in Davies 1904.
34. Lees E. 1868. *Botany of the Malvern Hills,* 3rd edn. London: David Bogue.
35. Godwin 1977, p.2.
36. Ibid., p.2.
37. R. H. Marks, p. 224 in Davies 1904.
38. L. Jacob, p.273 in Davies 1904.
39. F. J. Furnivall, pp.56–7 in Davies 1904.
40. C. P. Lucas, p.154 in Davies 1904.
41. Ibid., p.149.
42. Toynbee Hall was supported by groups from Oxford and Cambridge Universities. It offered residence to new graduates from Oxford and Cambridge who in return helped to educate the poorest men with a training for employment and life. Other teachers, like Tansley, were not resident.
43. Mullen 2009, pp.49–50.
44. Godwin 1958, p.1.
45. Kelly 1992, p.187.
46. Ayres 2005, p.55.
47. Murphy 1987, p.37.
48. Ibid., p.55.
49. Ibid., p.74.
50. Ibid., p.77.
51. Miranda Hill, Octavia's sister, helped form the Kyrle, a society 'to bring beauty home to the poor', in 1878. As well as decorating hospitals, schools, and clubs with frescoes and mottoes (with the enthusiastic support of William Morris), the Society's Open Spaces Committee worked to identify and protect open spaces in city centres. Octavia Hill was its Treasurer and Robert Hunter its legal advisor (Murphy 1987, p.58, p.66).
52. Darley 1990, p.310; cited from *More Air for London in Nineteenth Century* **23**, No. CXXXII, February 1888.
53. Sheail 1998, p.4.
54. Darley 1990, p.337.
55. Named after the Roman General, Quintus Fabius Maximus, who favoured attrition rather than head-on battle, Fabian socialists pursued gradual reform rather than revolution.

4 Highgate School, University College, London, and Trinity College, Cambridge

A slump in trade, followed by mass unemployment, then food riots in the centre of London, were the defining events of 1886 for the crowded masses of the city. But for 15-year-old Arthur Tansley the year was marked by a wholly gentler event, his move to Highgate School. On the northeastern edge of Hampstead Heath, and less than four miles from the city centre, Highgate was effectively in a different world. Noted for its nonconformist Unitarian and Quaker communities, its inhabitants thought of it as a village – as they still like to do today.

Arthur remained at Highgate School for two and a half years. Among his surviving letters and notebooks are two small volumes from his personal library, J. G. Baker's *Botanical Geography* (1875),[1] dated by hand on the inside cover '25th December 1886', and Herbert Spencer's *The Factors of Organic Evolution* (1887),[2] labelled on the inside cover 'AG Tansley 1888'. The first was one of the few English texts that considered possible reasons for observed distributions of plants, while the second was, in typical Spencerian style, a robust defence of Darwin's theories. On the subject of inheritance and evolutionary theory, Tansley was also reading *Studies in the Theory of Descent*,[3] an 1882 translation of August Weismann's German original. Another text read in translation was Julius von Sachs' *Lectures on the Physiology of Plants*,[4] which may have inspired his lifelong interest in physiology. (Soon after graduating from

Shaping Ecology: The Life of Arthur Tansley, First Edition. Peter Ayres.
© 2012 by John Wiley & Sons, Ltd. Published 2012 by John Wiley & Sons, Ltd.

Cambridge in 1894 he taught himself German in order to understand fully the nuances of German texts in their original forms.[5])

These books provide excellent pointers to the direction in which Tansley's career was to develop. Plant ecology would develop out of plant geography. Tansley would argue that ecologists should understand the underlying physiological processes that explained the distribution of plants and their peculiar form at a given location. And permeating his ecology was the concept of change, for he had realised that plant distribution, the type of vegetation, and the genetic composition of its component species, were all continuously changing.

The ready inference is that a highly talented schoolboy was being guided in his reading by one or more wise and perceptive teachers. But this seems to be far from the truth, for Tansley himself wrote of Highgate, 'the teaching of science – limited to perfunctory chemistry – was farcically inadequate'.[6] It remains unknown whether Tansley chose his biological reading himself or whether he was guided by someone outside the school. What is certain, for it comes again from Tansley himself, is that he asked his father if he 'could go to some place where he could get serious instruction in science and especially in biology'.[7] By now aware of his son's precocity, and on the advice of an unnamed friend, George made arrangements for Arthur to take the intermediate science lectures at University College, London (UCL), and there he remained from early 1889 to midsummer 1890.

University College was still relatively young and exciting. The first university to be established in England since the Middle Ages, UCL was founded in 1827 and in 1836 was incorporated, along with King's College, as the University of London. UCL holds a special place in the history of English education because it broke the stranglehold of the old universities, Oxford and Cambridge. A large proportion of the new urban population belonged to nonconformist churches and were, therefore, barred from those ancient universities which, until the Universities Terms Act of 1871, demanded that all their students who sat for degrees should subscribe to the Thirty-Nine Articles of the Anglican faith. (Printed at the back of the *Common Book of Prayer*, the Articles have formed the basis of Church of England doctrine since earliest times.) For centuries, nonconformists could in consequence study at only the four Scottish Universities, St Andrews, Glasgow, Aberdeen, or Edinburgh, if they wished to take a degree.

When teaching at UCL began in 1828, the first courses were in the arts, law, and medicine, thereby going beyond the traditional subject areas of moral and political philosophy to 'those sciences which consist in the examination of the laws and properties of material objects'.[8] The founders of the University, who included in their number Quakers, Baptists, Jews, and Evangelicals, restricted teaching to such secular subjects as a matter of principle. UCL quickly became known as the 'Godless Institution in Gower Street' and was not without its opponents:

> Ye Dons and ye Doctors, ye Provosts and Proctors,
> Who are paid to monopolise knowledge,
> Come make opposition by voice and petition
> To the Radical Infidel College.
>
> T. E. Hook, *The Cockney College*[9]

Not everything went smoothly for the young institution. Among the first pro-fessorial appointments were J. F. Meckel to the Chair of Morbid and Comparative Anatomy and William J. Hooker (father of Joseph Dalton Hooker) to the Chair of Botany and Vegetable Physiology. Neither took up his duties because UCL could not guarantee them sufficient income; professor's income was related to the numbers of students enrolled, and that was uncer-tain in a new institution. In Hooker's case there was an additional reason for withdrawal, which was the College's decision that it could not afford to estab-lish a botanic garden.[10] The post was filled by John Lindley, whose books, such as *Introduction to the Natural System of Botany*,[11] promoted the new 'natural' system of plant classification which deviated from the rigid 'sexual' system of Linneaus by introducing morphological features and the positional relation-ships of organs. An accomplished administrator (whose great achievement was to chair the Parliamentary committee that 'saved' Kew Gardens from Treasury cuts),[12] Lindley was an expert on orchids and, like Daniel Oliver, the man who succeeded him at UCL 21 years later, a systematist. Neither man involved himself with either plant function or distribution, save where these related to the uses of plants in medicine, for most of their lectures were designed for students studying medicine. Daniel Oliver is notable, however, for one innovation: he confronted his students with real plants. Before each of his lectures, always at 8:00 am, he laid out six novel specimens which his students must learn to recognise, 'a quarter of an hour being taken from the lecture for this purpose'.[13] Oliver gave his lectures early in the morning because he then had to return to Kew - where he lived - to complete a full day's work as Keeper of the Herbarium.

As explained by his son, Francis Wall (Frank) Oliver (see Figure 1.3), who succeeded him at UCL in 1888, the College's Chair of Botany made few demands upon its first occupants.

> The professor only delivered his lectures and departed; but for the accident that, here and there, the seed fell on good ground, the progress of the subject was not directly advanced. The professor held his college post primarily as an additional source of income, and there were no facilities for developing the science within the precincts. There were few positions of emolument open to a botanist, and of the deliberate training of advanced and postgraduate students, there was none.[14]

There were small improvements in the physical provision for teaching botany at UCL when in 1880 new building at last provided a laboratory, making it possible to introduce practical tests into the various examinations, although in these respects botany lagged behind zoology, as explained later. Teaching was galvanised and a tradition for research of the highest order was established only after Frank Oliver was appointed lecturer in 1888. (He gained complete control over teaching when he was promoted to a full professorship two years later.) Tansley was inspired by Oliver, declaring that 'in a year and a half [I] certainly learned more both theoretically and practically than [I] ever did at any other course in [my] student days'.[15] Oliver lectured slowly, with pauses enabling his students to keep up with their note taking, but when the next sentence came 'it was always perfectly lucid and to the point'. His practical classes were based around microscope work and were 'hands on'. Following the precepts of T. H. Huxley, students were expected to cut, stain, and mount their own sections of plant material.[16]

Frank Oliver was not the only outstanding scientist to whom Tansley was exposed during those first 18 months at UCL, for also teaching at the College were William Ramsey and E. Ray Lankester.

After leaving his native Scotland, Ramsey had followed a path well trodden by the best postgraduate chemists and biologists of his day – he had gone to study in a German laboratory. The significance of the German laboratories was that they placed great emphasis on experimentation, and Ramsey proved to be an exceptional experimentalist, winning the Nobel Prize for chemistry in 1904. (He rolled his own cigarettes, claiming that machine-made ones were unworthy of an experimentalist.) Appointed to UCL's Chair of Inorganic Chemistry in 1887, Ramsey is perhaps best known for his discovery of the family of 'inert gases', comprising argon, helium, neon, krypton, and xenon, each announced between 1885 and 1890.[17] Unfortunately for Ramsey's students, although his lectures were good, and occasionally inspiring, his practical classes were not. He could not translate the excitement of research into his teaching, leaving Tansley with the memory of interminably 'pouring one reagent into a test tube containing another and noting the formation, texture and colour, or the absence of formation, of a precipitate'.[18]

Lankester was a protégé of T. H. Huxley, his speciality, like Huxley's, the comparative anatomy of invertebrates. Like Huxley he was a convinced materialist. He had known Huxley since childhood days when the renown of his father, Edwin Lankester – the unused 'E' in Ray's name stood for Edwin– as a doctor-naturalist had forged a close friendship between Edwin and Huxley. (The Lankesters were exceptionally well connected, Charles Darwin and William Hooker being numbered among their family friends.) Ray Lankester studied under his father's friend, Huxley, at Oxford before taking his MA. Later he helped Huxley teach his ground-breaking classes at the Science School, South Kensington. The government had set up the School in 1872, under Huxley's direction, with the express purpose of training young men to teach science. What was ground-breaking was that Huxley had insisted that laboratories should be included in the School, whereby students could gain that first-hand, practical experience of the anatomy of animals and plants which he deemed essential. Such experience, he believed, would give them the confidence to challenge the accepted wisdom enshrined in textbooks. Lankester carried away from the Science School this novel spirit of learning through practical experiment and, when he was appointed to the Chair of Zoology at UCL in 1875, he immediately set about introducing it to his students (in doing so, he set an example his botanical colleague, Frank Oliver, would follow).

Lankester was an exceptionally gifted lecturer who 'illustrated his expositions by rapid and admirable drawings in coloured chalks on the blackboard'.[19] His enthusiasm was so great that his lectures regularly spilled over into the lunch hour, leaving students with only a few minutes in which to find some lunch before the afternoon's practical classes began. It was Tansley who, assuming what was to become his familiar mantle of leadership, collected a petition, signed by most of his fellow students, emphasising how much the class enjoyed Lankester's lectures but respectfully asking him to allow them the proper time for lunch. For a few days the situation improved, but Lankester soon relapsed into his old habits.[20]

Lankester differed from Huxley in two important ways. Whereas Huxley was content to have his students examine dead material – some critics have subsequently called him a 'necrologist' – Lankester insisted on his students seeing living material, as far as possible. It was he who introduced the term 'bionomics' to denote the study of the living animal *in its environment*, anticipating a basic method of ecology. And, whereas Huxley was on most occasions a diplomatic advocate of evolutionary theory, Lankester tended to let his emotions rule his head. He enjoyed nothing better than a public scrap, such pugnacity extending beyond his defence of evolutionary theory. Thus, in addressing the British Association for the Advancement of Science (BAAS) in 1883, on the subject of 'Biology and the State', he publicly denounced the government for its failure to support science in Britain. He compared Britain with Germany – of which he had first-hand knowledge, having studied at Jena and Leipzig – pointing out that per capita there were roughly five times as many salaried posts for scientists in Germany than in Britain.[21] Many years later, in the spring of 1914, when Britain was sliding towards war with Germany, he risked further unpopularity by publishing in *Nature* an article pointing out that the German government had set up a department of state to assist in the protection of nature and had already created over 100 reserves.[22] While attempting to support the newly formed Society for the Promotion of Nature Reserves (SPNR), he was hoping to provoke the government into action to protect wildlife. In the latter aim, he was unsuccessful.

Lankester was a large, rumbustious, and colourful character, whose presence must have been inescapable within UCL. His views were well publicised in academic circles outside the College and he attained some celebrity when Arthur Conan Doyle chose him as the model for the evolutionary biologist, Professor Challenger, in his novel, *The Lost World* (1912). Challenger even refers to him by name when he mentions 'an excellent monograph by my gifted friend Ray Lankester!'.[23]

Tansley freely admitted to the brilliant style of Lankester's teaching but did not dwell upon what might be expected to have been among the most exciting elements in its substance, namely Lankester's far-sightedness in recognising the importance of nutrient cycling in the environment and his concerns about the extinction of plant and animal species as a result of man-made pollutants. Could it be that in the mid-20th century – when Tansley was looking back at the formative influences upon his career, and when Marxism was an anathema to most people outside the Soviet bloc – the origins of what he had learned from Lankester were best left undisturbed? For Lankester had been not simply an ardent socialist – his father, Edwin, had been a supporter of Chartism – he had been a personal friend and admirer of Karl Marx. (He was one of only 11 mourners at Marx's funeral in Highgate Cemetery in 1883.) Lankester had read Marx's *Capital: a Critique of Political Economy* in its German original, 'with the greatest pleasure and profit'.[24] In that book and in his essay, 'The Effacement of Nature by Man', Marx had argued that capitalism had caused an irreparable rift in what he called the 'metabolic interaction' between human beings and the earth. Quite simply, food was produced in the countryside (at the expense of soil nutrients), transported to the cities where it was eaten by the human population, and the nutrient-rich waste products, human excretia, were then allowed to flow into sewers and rivers. Nutrients were wasted, rather than being returned to the land as they had always been in the pre-industrial age.

Lankester's political views were no secret, nor were his views on the ultimate usefulness of science. In his 1883 BAAS address he said:

> Through [science] we believe that man will be saved from misery and degradation, not merely acquiring new material powers, but learning to use and to guide life with understanding.[25]

Tansley was privileged to have enjoyed the teaching of a man whose mind overflowed with challenging ideas. It was in Lankester's forthright nature to accumulate enemies but he also inspired many young men who were destined to become leading biologists. Two such, Charles Elton and Julian Huxley (T. H. Huxley's grandson), would prove respected and valuable collaborators of Arthur Tansley.

***

In October 1890 Tansley went up to Trinity College, Cambridge. Student societies and gatherings dominated his first year but in his second he knuckled down to study and was rewarded with a College Exhibition, which meant he received a small allowance – though less than that of a Scholarship – and certain privileges within Trinity.[26] In his second year he added geology to his other subjects, which brought with it the chance to join field trips, including two-week-long ones held in the summer and led by Professor Thomas McKenny Hughes. Tansley could not only learn geology at first hand from a gifted teacher but he could also, in theory, enjoy the company of female students from Newnham and Girton Colleges. It seems that, in practice, there was not much time for flirtation because the professor worked his students hard and his wife, 'a wise and excellent chaperone', always accompanied the mixed parties.[27]

The bare facts of Tansley's time at Trinity are that in 1893 he was placed in Class I in Part I of the Natural Sciences Tripos, for which he read botany, zoology, physiology, and geology, and in 1894 was awarded a first class honours degree after reading botany in Part II. Tansley said little about the teaching he had received in those Cambridge years. Within the Botany School, there was certainly no inspiring teacher to match Oliver at UCL. Tansley was indeed unlucky in being an undergraduate during a period when the School was struggling. Its professor, Charles Babington, was old, ill, and rarely seen. Sydney Vines, who had done so much in the early 1880 s to reorganise teaching along German lines, introducing a little physiology and some practical work into the course, had departed to take the Sherardian Chair in Oxford. This left Francis Darwin as, effectively, Head of School, and having to cope with poor facilities and disaffected colleagues. A heavy teaching load fell upon Darwin, who was simultaneously having to balance the demands of school administration, his own successful research career in plant physiology, and his duties as his father's biographer.[28] It is little wonder then that there is no record of Tansley ever having admitted to being excited by his Cambridge teachers. The Botany School revived only in 1895 – a year after Tansley's departure – when Babington died and, with Darwin's best wishes, Harry Marshall Ward was appointed to the chair.

The undergraduate experience is, of course, about more than classrooms and laboratories. Tansley appears to have settled easily into undergraduate life. Writing to his mother on 6 November 1890, and in spite of having suffered a bad cold and protracted dental treatment, he cheerily tells her how he has joined the Cambridge University Musical Club, which performed chamber music on Saturday evenings, and had long discussions with his fellow undergraduate, Bertrand Russell,[29] about Shakespeare's tragedies. There is a footnote to this letter which is significant in terms of what would become a major interest in his life – psychology. He writes, 'Went to a meeting of the Psychical Society last night … and heard Mr Myers[30] discourse on "subliminal self"'.[31]

Tansley had made Russell's acquaintance in his first year at Trinity but it was in his second that their friendship truly blossomed. Early in the Michaelmas term he wrote to his mother:

> I like Russell the more I see of him. I think he has the finest and subtlest _feeling_ for first-rate qualities of all kinds of any one I have ever met, joined to a splendid intellect … [adding, more prosaically] please let me have directly Romanes' 'Scientific Evidences of Organic Evolution'[32] [which he had left at home] …
>   *Arthur to Amelia Tansley, 23 October 1891, from Trinity College, Cambridge*[33]

Soon the two young men were collaborating in the production of a student newspaper. In a letter dated 1 May 1892 to his grandmother, Russell wrote:

> We have been quite ready [?] getting together the first number of the Cambridge Observer … Yesterday afternoon I went to town with [Charles Percy] Sanger and Tansley to see Ibsen's Doll's House, ostensibly for the purpose of criticising it in our paper …[34]

The *Cambridge Observer* ran to 21 issues spread over three terms. Like most student newspapers it seems to have been a mixture of gossip and news. As their friendship developed, the pair chose to spend part of their vacations together. In a letter, almost certainly from 1892 but simply headed 'Mar 24', Tansley wrote to his mother from 'Albergo di Nettuno, Pisa'. Holidaying in Italy, he told her how he and Russell had travelled through the south of France, stopping at Marseilles, the Riviera, San Remo, and Spezia, from where they were able to visit the house of one of Russell's favourite poets, the late Percy Shelley. Next they would go to Florence.

One of Russell's circle who became Tansley's lifelong friend was George Macaulay Trevelyan, the future historian and Master of Trinity College, teacher at the Working Men's College (WMC) from 1899,[35] and distant relative of the Darwins and Huxleys. When George entered Trinity in 1893, he was joining in the college his brothers Charles[36] and Robert, who had entered in the previous years. Among this undergraduate group there were the 'usual interminable discussions on the universe – on philosophy, psychology, religion, politics, art and sex'.[37] The group's political sympathies were almost certainly left of centre. Socialism was one of the great causes in Russell's life and George's father, Sir Otto Trevelyan, and George's brother, Charles, were stalwart liberals – though Charles moved leftward, joining the Fabian socialists later in life. A love shared by Tansley, Russell, George Trevelyan, and Sanger too, was vigorous country walking, although George and Charles Trevelyan could outstrip their friends by some distance; George's guests found his idea of a short stroll after lunch was a 30-mile hike.[38]

Cambridge, and Trinity in particular, wielded at this time enormous political influence. In 1893, sixty-eight of the 105 Cambridge-educated members of parliament were from Trinity, while a third of W. E. Gladstone's cabinet were Trinity men.[39] People like the Russells and the Trevelyans, who belonged to the land-owning aristocracy, were commonplace in Trinity. Although the Tansleys were financially comfortable, their modest wealth came from what was disparagingly called, in the parlance of the day, 'trade'. It is testimony to Tansley's intellect and self-confidence, as well as to the best liberal instincts of Russell and the Trevelyans, that in the class-ridden 1890 s he was able to integrate happily with them. If his parents had not always paid their son sufficient attention, they had clearly taught him the invaluable lesson that he, like them, could be comfortable in the company of the brightest, most influential men and women of the day – in a world where ideas mattered above all else.

Notes

1. Baker LG. 1875. *Elementary Lessons in Botanical Geography*. London: L. Reeve.
2. Spencer H. 1887. *The Factors of Organic Evolution*. London: Williams & Norgate.
3. Sachs J von. 1882. *Vorlesungen über Pflanzenphysiologie*. Würtzburg. (English translation by H. Marshall Ward. 1887. *Lectures on the Physiology of Plants*. Oxford: Clarendon Press.)
4. Weismann FLA. 1875–1876. *Studien zur Descendztheorie*, 2 vols. Leipzig: Wilhelm Engelmann. (English translation by R. Meldola. 1882. *Studies in the Theory of Descent*, 2 vols. London: Sampson Low, Marston, Searle & Rivington.)
5. Godwin 1977, p.4.
6. Ibid., p.3.
7. Ibid., p.3.
8. Taylor 1968, p.11.
9. Ibid., p.11.
10. Hale Bellott 1929, p.38.
11. Lindley J. 1830. *Introduction to the Natural System of Botany*. London: Longman, Rees, Orme, Brown & Green.
12. After exposing the incompetence and extravagance which had led the Treasury to threaten closure, Lindley's committee recommended in 1839 that the Royal Gardens, Kew, should be made over to the nation and should become the headquarters

of botanical science for England, its Colonies and Dependencies (F. Keeble, p.169 in Oliver 1913).

13. Anon. 1927. *An Outline of the History of the Botanical Department of the University College, London.* London: University College, p.10.
14. F. W. Oliver, cited by Hale Bellott 1929, p.393.
15. Godwin 1977, p.3.
16. Godwin 1957, p.228.
17. www.ucl.ac.uk/ramsey-trust/life.
18. Godwin 1957, p.228.
19. Ibid., p.227.
20. Ibid., p.228.
21. Lester 1995, p.162.
22. Sheail 1976, p.197.
23. Doyle 1912, p.30 of 2008 reprint.
24. Foster 2000, p.222–3.
25. Cited by Lester 1995, p.162.
26. Godwin 1957, p.229.
27. Ibid., p.229.
28. Ayres 2008, p.134.
29. Bertrand (Arthur William, 3rd Earl) Russell. Philosopher, mathematician, and winner of the Nobel Prize for literature, 1950.
30. Frederic Myers, Cambridge classicist and prominent member of the Society for Psychical Research.
31. Branscoll (a collection of family letters found at Branscombe, South Devon).
32. The book, or extended essay, was published in 1882. The youngest of Charles Darwin's academic friends, the physiologist George J. Romanes is credited with inventing the term 'neo-Darwinism'.
33. Branscoll.
34. Griffin 1992, p.9.
35. Cannadine 1992, p.62, p.144.
36. As the eldest son of Sir George Otto Trevelyan, Charles inherited the Baronetcy of Wallington, a large estate in Northumberland (now Northumbria). Charles and his brother, George, helped found the Youth Hostel Association in Britain in the 1930s, the first hostel being at Wallington Hall. See also footnote 39 to Chapter 12.
37. Cameron 1999, p.6.
38. Cannadine 1992, p.147.
39. Ibid., p.20.

5 Teaching at University College, the Chicks, and Marriage to Edith

Winning a first class honours degree in the finals examinations of 1894 was a remarkable achievement for Arthur Tansley, not because there was any suggestion that the brilliant young man did not deserve it, but because throughout his last year at Cambridge he had been leading a double life. The previous summer, Frank Oliver had offered him a teaching assistantship at University College, London (UCL). The salary was £50 and he accepted the job, confident that he could both teach and study successfully – as he did. Tansley remained at UCL until 1906, according to the College's records progressing from Quain student, 1893–1898, to Assistant Professor of Botany, 1895–1903 (the positions and dates overlap), and, finally, to lecturer in 1904–1906. Whichever his title, in practice he was Oliver's assistant, gradually taking more and more teaching from Oliver's shoulders and, as he gained experience, helping Oliver develop the botanical curriculum.

> During five years I demonstrated regularly with the professor to the large elementary class. I also undertook a large share of the advanced lectures, dealing with among other topics, Anatomy with Reproduction, Floral Biology, the Algae, and the Morphology of Flowering Plants. In 1899 I delivered a comprehensive course on the Morphology of the Vascular System of Plants, by far the fullest treatment of this subject that had been attempted.[1]

Shaping Ecology: The Life of Arthur Tansley, First Edition. Peter Ayres.
© 2012 by John Wiley & Sons, Ltd. Published 2012 by John Wiley & Sons, Ltd.

Perhaps the most significant contribution to curriculum development made by Oliver and Tansley was the introduction of field excursions in 1903–1904, notably including residential courses, through which undergraduates could make an intensive study of vegetation types carefully selected for them. Those courses are the subject of the next chapter, the focus in this being Tansley's personal development. Undoubtedly, teaching was demanding, especially the field work which required a considerable amount of preparatory surveying of potential sites and identification of their commonest plants, but plenty of free time remained when an energetic man in his twenties, as yet unburdened by family responsibilities, could enjoy himself.

Tansley was thus able to keep up with friends he had made at Cambridge. Among them were the Blackman brothers, Vernon and Fritz (Frederick Frost, or 'F.F.') (see Figure 1.3). They were fellow botanists whose lives became closely intertwined with Tansley's both professionally and socially, Fritz becoming Tansley's brother-in-law. Vernon entered Cambridge in 1892, at first sharing rooms in St John's College with Fritz and a third brother, Sidney.[2] After Vernon graduated and was appointed as an assistant in the British Museum (Natural History) in 1896, he and Tansley shared an apartment in London.[3] When his friend applied for a Fellowship at St John's, Tansley spent three days helping Vernon perfect 'his great paper', telling his sister, Maud, 'I got nearly as excited about it as if it had been my own'.[4] The two young men shared many 'likes', including what they judged to be an excellent vegetarian restaurant near their apartment. For a brief period they became dedicated vegetarians. Although Vernon found time to give lectures on 'Vegetable Physiology', at UCL, his principal interest was in mycology (fungi). To assist with displays and teaching he employed the skills of a maker of wax models, Miss Edith Emmett, his future wife. They met because she had previously been making models for E. Ray Lankester and was sent to Vernon for scientific advice and help. In 1907 Vernon was appointed the first Professor of Botany in the new University of Leeds, moving back to London in 1911 when he took a chair at Imperial College.

The Blackmans, like Tansley, were Londoners, though in their case they came from a poorer region, Lambeth, south of the Thames, where their father practised medicine and their mother was a medical superintendent in a prison. The parents can have had little time for their children. Thrown upon their own devices, the brothers occupied themselves with reading and walking. Some of their exploits were extraordinary: the brothers would sometimes set out late in the evening and walk through the night into the nearest countryside, returning in the early morning when the streets were being washed and the markets were waking up.[5] Both started studying medicine, but both broke with the habit of their family and turned to botany; it is a moot point whether, as some would have it,[6] they had been inspired in childhood by reading the 12 volumes of Sowerby's *English Botany* in their father's library.

In these years, Tansley was able to keep up too with Bertand Russell and his circle. In a letter dated 18 September 1896, he tells Maud:

> I have been staying at Haslemere for a few days … in a cottage belonging to Russell's father-in-law, Mr Pearsall-Smith.[7] Russell was at Friday's Hill, Mr Smith's house, part of the time, and several Cambridge men of our time, Robert Trevelyan, Fountain, and Vaughan-Williams[8] came down at different times, so you can imagine I had a splendid time.[9]

In the same year, 1896, he found time to go to the Wagner Festival in Bayreuth with his mother (they may also have visited Bayreuth in the previous year, but the record is ambiguous).[10] Life away from UCL was not all play; among other things, he prepared a lengthy essay on 'Natural selection considered as a special example of the general principle of evolution', which he submitted to the annual competition for an Arnold Gerstenberg studentship at Cambridge. These studentships are offered by the University to this day. Tenable for one year in the first instance, they are open to those who have graduated in natural science and intend to pursue a course of philosophical study. He shared the prize jointly with Charles Samuel Myers,[11] although the use to which he put it is not clear.

Reflecting his growing knowledge of algae, he helped Oliver update a section on the Gamophyceae – simple green algae – in *The Natural History of Plants* (= *Pflanzenleben*) (1896), a massive German text of Kerner von Marilaun's that Oliver was helping to translate into English and to revise. In 1896, and on Oliver's recommendation, he helped the elderly Herbert Spencer revise his *Principles of Biology* (first published in 1864), a task which, as explained in Chapter 12, gave him particular pleasure. In 1899, Tansley gave a university extension course on ecological plant geography to the students of Toynbee Hall, an indicator not just of his social conscience but also of his changing interests in botany.[12] Professionally, it was a period of novelty, excitement, and burgeoning confidence; personally, it was a significant period because it was when he met Edith Chick, who was to be his wife.

***

In 1878, UCL absorbed the London Ladies' Educational Association and formally admitted women to degree courses and examinations – the first British university to do so – although for several years already they had been able to attend certain lecture courses. Three hundred and nine women were admitted, approximately half the number of men already studying.[13] A little more than ten years later, Edith Chick and her sisters were to benefit from the enlightened policies of UCL.

Edith's parents, Samuel and Emma (née Hooley), were married in 1867. Emma immediately moved into the terraced house at 5 Newman Street (just west of Tottenham Court Road and on the fringe of Bloomsbury), the address at which Samuel had been living for the last four years, and the ground floor of which was a showroom for his family's lace-dealing business. The traditional centre of English lace making was East Devon and one small town, Honiton, had given its name to the particular type of lace in which the Chicks specialised – all of Queen Victoria's daughters were married in dresses trimmed with Honiton lace. However, fashions were changing.[14] Recognising that the future lay in the highest quality lace, either English or foreign, and preferably antique, Samuel took his business to London where he was better placed to wheel and deal, and where richer, more discerning, customers could be found.

Samuel also did what every modern business manual would advise; he diversified. He bought and sold commercial properties, shrewdly confining his dealings to the small area around Newman Street, the area he knew best. Samuel secured thereby not only the future of the family lace business but, with his wider dealings, the Chicks' future prosperity.

Between the years 1868 and 1887 Emma bore Samuel 12 children, all born at Newman Street. The first born were twins but the little girl died in infancy, so when Edith arrived in 1869 she was the second child of the family and, in time, would be the oldest of seven girls.[15] Five of the girls graduated from the University of London, with Edith, Harriette, Frances, and Dorothy all studying either science or medicine,[16] a remarkable statistic in any age but particularly so when educational opportunities for women were limited. Their mother's behaviour may, by accident rather than design, have encouraged these obviously talented young women to take such an unusual path for, although Emma loved tiny babies, it seems that once their babyhood was over she quickly turned her time and attentions to the next. The impression gained from the family's history, written by Margaret Tomlinson (Tansley's second daughter), is of parents who were devoted to each other, who brought up their children to pray twice daily, to visit a nonconformist chapel twice on Sunday, and to regularly attend Sunday School, but who did not show to their children the warmth that they displayed to each other.[17] The younger girls were indeed lucky in having such a forceful and determined elder sister as Edith, for it was she who blazed the trail leading to a university education.

The Chick girls started their formal education at Miss Harrison's private day school in Gower Street (see Figure 3.1), which took boys up to the age of eight and girls up to the age of 16 years. The small school attracted children from the well-to-do families of Bloomsbury, many professional and some academic.[18] Lucy Harrison's love of the classics and poetry were strongly reflected in what was taught to the girls. A Yorkshire woman from a Quaker family, she had a natural authority. There were no rules at Gower Street for, in Miss Harrison's presence, dissent would be shameful, the 'inexpressible charm of her presence' making it impossible.[19] Above all, the school encouraged any talents the girls possessed.

One of Edith Chick's contemporaries at Gower Street was Ethel Oliver, daughter of Professor Daniel Oliver and brother of Frank. The Olivers were a Quaker family – one probable reason why Ethel was placed at Gower Street, the other being that Daniel travelled regularly from Kew to UCL to give his classes. Ethel was known by another friend, the future Sapphic poet Charlotte Mew, as peaceful and Quakerish (not unlike her brother). By contrast, Charlotte's spirit was far from peaceful. Ethel, the Chick girls, and Charlotte were close friends. The latter stayed for extended periods with both the Olivers and the Chicks. Friendships made at Gower Street lasted into adulthood. Charlotte described in her diary one great adventure in 1901 when 'six unmated females', all thirty-ish, holidayed together at the convent of St Gildas du Rhuys in southern Brittany. The composition of the party is not entirely clear but, almost certainly, it included Edith, Harriette, Margaret, and either Mary or Elsie Chick.[20]

Samuel Chick's plan had been for Edith to enter the family business but when, at 16, the time came for her to leave Gower Street, Edith fought against being incarcerated in the Newman Street showroom, declaring that she wanted to become a teacher. With commendable open mindedness, and utilising a connection he had made through the Marylebone Liberal Association, of which he was Honorary Secretary, Samuel consulted Lady Stanley of Alderley, a pioneer in the cause of higher education for women and a founder member of the Girls Public Day School Company. The story handed down in the Chick family is that in spite of patronising Samuel, because he was 'in trade', Lady Stanley encouraged him to send Edith to Notting Hill High School, one of two schools founded by the Company some 13 years earlier. 'Edith entered the school in 1886 and, despite her late arrival, never looked back. She particularly enjoyed her first introduction to algebra and to science – subjects rarely taught to girls at the time. The Chick sisters were lucky, for they were taught by the first wave of female "graduates" emerging from Oxford, Cambridge, and London universities; almost certainly their education was more advanced in academic terms than that received by their brothers.'[21]

From Notting Hill High School, Edith moved to UCL in 1889. In her first year she studied physics and mathematics, to which was added in the second year, botany – almost certainly taught under the direction of Frank Oliver. At this point, she was probably not a classmate of her future husband, Arthur Tansley, for he went up to Cambridge in 1890, but she may have met him during other science classes, or socially, during her first year. By her fourth year, however, when Edith was specialising in botany, Tansley was back at UCL helping

Oliver, so it is highly likely that their paths crossed, and frequently so. On Monday 31 December 1894, *The Times* newspaper carried a short announcement to the effect that Edith Chick had graduated with an honours degree in botany, second class. She was one of only four women in the list for all subjects.

It is not entirely clear what Edith was doing during the years 1895 to 1898. It seems that, either through her own efforts or more probably thanks to Samuel's support, she was paying for further classes in botany. At the end of this period, in 1899, she was the first female ever to be made a Quain student, receiving an annual grant of £100. There is evidence that she was also active in the students' union, for the minutes of the Women's Union Society of 6 February 1899 record there was a discussion of Miss Chick's proposal 'that the officers of the various societies belonging to the men students and the Junior Staff should be invited to the social evening'.[22] Edith's proposal was carried unanimously.

Among those who would have been invited to this annual event was Arthur Tansley (from this time onwards known to his close friends as 'A.G.'), who obviously had an eye for the ladies. Contributing an article about Girton girls to the UCL student newspaper, the *Gazette*, Tansley recommended a trip to Cambridge where 'Girton feet and ankles are proverbial … and hockey skirts are short'.[23]

On his return to UCL, and as one of its youngest members of staff, Tansley had agreed to be appointed Vice President of the Committee of the Students' Union. In 1898 he took over the Presidency, for two years, and was then its Treasurer for a further two years. The formation of such a union by students presented the College with an unprecedented threat to its authority, which was why for a number of years it insisted that the senior posts should be held by members of staff. Sensibly, it chose younger members of staff for the task. Tansley was young enough to share the students' interests and pastimes but, as a member of staff, sensible enough to exercise responsibility. Thus, he was sometimes called upon to moderate student opinions, as in presenting to the authorities their views on 'Irish stew', 'French pastry', and the inadequacies of the Refreshment Room.[24] Responsibility was rewarded with perks. Not least, greater opportunities to meet the opposite sex, something that Tansley clearly enjoyed. In reporting one social event in June 1901, the *Gazette* reads,

> Dr Gregory Foster and Mr Tansley both were present, each asking the other between the dances whether he was 'ready to go yet'. The fascinations proving too strong, however, they both danced to the very last extra.[25]

Tansley once showed similar energy on the running track. He used to recall with amusement how in his first year back at UCL – when he was also completing the Cambridge Tripos – he attended the staff sports at UCL in the customary morning dress and, thus attired and carrying his top hat, he won the 100 yards race.[26]

The Quain studentship held by Edith from 1899 was the same one held earlier by Tansley, before he was made Assistant Professor in 1895, and was allied to the Quain professorship held by Frank Oliver. Edith was effectively Oliver's assistant, helping him among other things with his research on fossil botany. It was during this period that she became professionally, and personally, involved with Tansley. By September 1900 she was checking through the first proofs of a joint paper, 'Notes on the conducting tissue-system in Bryophyta', which they were to publish in the *Annals of Botany* in 1901. Assuming a minimum gestation period of two years from the start of the research to the submission of the paper, its refereeing, acceptance, and publication, she could easily have been working with Tansley *before* the studentship commenced. For his part, Tansley had already published in 1896 two papers in *Science Progress* (his first professional publications) on 'Stelar theory' – the evolution of different architectures of conducting tissues in different groups of the plant kingdom. Now, with Edith's assistance, he was able to develop in depth his theories with respect to one group, the Bryophyta (mosses).

Edith's letter to Tansley, written from her family home in the small South Devon village of Branscombe on the wet afternoon of Sunday 2 September 1900, reveals not only a young woman's frustration with stuffy village life, but a familiarity with the man who was her supervisor. She was returning to him proofs of their paper, thus the letter begins in a business-like manner with comments such as

> Is 'contents' singular or plural?
> In the description of the leaf traces and the way they enter the central hydroid cylinder etc shouldn't there be reference to a figure somewhere and why are half the leaf traces numbered with roman figures?

Soon, however, its tone changes to one that is a mixture of admiration and tongue-in-cheek humour:

> I am astounded to find how interesting the paper is. I think the conclusions are so jolly and I am sure I never appreciated the importance of the strand of Diplophyllum and P. Hibernica ... before I saw the awed and inspired way in which you approached them.
> *Edith Chick to Arthur Tansley, 2 September 1900, Branscombe*[27]

Having heard that he is going to the Annual Meeting of the British Association, 'to hold forth on the Bryophytes at Bradford', she teases him, 'shall I be able to seen an account of it in *The Standard?* That's the paper we see in this stronghold of Church and State'. Edith's letter ends, 'When I think that in a month's time you will see a tropical rain forest, I am consumed with envy'. Tansley was to depart in the autumn of 1900 on a lengthy botanical excursion through Ceylon, Malaya, and Egypt. Their relationship had clearly become even closer by the time he left on his travels, for not only was she allowed to correct the final proofs of their paper in his absence but they shared confidences, her letters keeping him in touch with news of the people and, more importantly, the latest gossip at UCL. One of Frank Oliver's demonstrators was the butt of their humour:

> I find I was hasty in likening Mr Worsdell unto a High Churchman, he is more of the Earnest Rivivalist – There's a 'come, be saved attitude' ... He appealed passionately last time to all those of us who had hitherto led a careless (botanical) life who had given no thought to the morphology of the ligule [a small membranous outgrowth found at the inward facing junction of the leaf sheath and blade in grasses] ... Occasionally he makes lists on the board of those botanists you may not trust on any subject, generally a long imposing one, then in opposition those few to whom light has been given, few indeed, and always headed by Celakovsky.[28] Once in a moment of passion he brought out a postcard from the Prophet himself and after gazing rapturously at it put it back again, next to his heart.
>
> *Edith Chick to Arthur Tansley, Sunday 10 February 1901, Branscombe*[29]

Edith encouraged Arthur to keep a diary during his travels so that she could share his experiences of the tropics. By 1903 the pair were publishing again in the *Annals of Botany*, this time 'On the structure of *Schizaea malacanna*' (a fern). And this time Edith was the senior, first named, author, showing she had developed more independence of thought and had done most of the work. Ironically, the paper proved a landmark in Edith's life, signaling as it did the end of any career she might have had as an independent researcher, for on Thursday 30 July 1903 she married her erstwhile supervisor and collaborator. She was to accompany her husband on countless botanical excursions and expeditions but was thereafter always a hostess, a coordinator, a secretary; in short, an adjunct to *his* career.

***

Edith and Arthur's wedding was in the small church at Branscombe, near Sidmouth and the Chick family's roots. Arthur may have gone against the wishes of his nonconformist father-in-law by marrying in church rather than chapel.

The tiny coastal village of Branscombe, hidden deep in a fold of the Devon countryside, was a second home for the Chicks. Samuel was in the habit of renting a large holiday home there each summer, where they and their friends could escape from London and the rigors of city life. Thanks to the survival of a few pages from the diary of Edith's sister, Harriette, a little of the background to the wedding is known. Harriette was the next oldest of the Chick girls, some six years younger than Edith, and the sister destined to be the most academically distinguished. (From botany she turned to bacteriology – she was the first woman to work for the famous Lister Institute – and then to nutritional studies. For her work on the deficiency disease, rickets, she was in 1944 made a Dame of the British Empire.) Immediately before the wedding, Harriette returned from Germany to London where she seems to have shared a house with friends and, possibly, family. Her diary reads:

> **Sunday 26.7.1903** Got home from Germany to find the house much excited … Inspected the wedding presents and the 14 serviette rings and in the afternoon Stopey [Marie Stopes][30] and Norman and Lot [Charlotte Mew, who was especially fond of Harriette] came to tea. Stopey was rather sweet on the subject of the D.Sc. amusingly so. Norman finally went off at about seven and Lot stayed. E's clothes etc the absorbing subject of conversation.
>
> **Monday 27.7.1903** They went shopping. At 3pm 'A select party' saw the bride off, including Aunty Smith, Lot … [the bride was going to Branscombe ahead of Harriette and the main party]
>
> **Wednesday 29.7.1903** Showery day again, but we picnicked in true Chickian way [the Chicks were notoriously frugal] on Colyton moors whilst we got heather to decorate the house, ate bread and cheese, got flowers, sat in the road and behaved in a most unbecoming manner – especially the bride. Weather prospects awful.
>
> **Thursday 30.7.1903** Day broke looking promising. Bride has been sleeping most disgracefully well. Can't realize there is to be a wedding today. It seems quite unlike it somehow. Entertaining Mrs Cane [?], getting the brides boots heeled and fetched and decorating the house with flowers took up all of the morning. The village is en fete, banners, flags … an arch.

After a brief visit to Ireland, during which Edith was seasick and generally unwell, the main honeymoon was delayed until early next year, when the couple visited the south of France, followed by Italy, so escaping the worst of the British winter. They were rewarded with warm sunshine, at least in Cannes, where they stayed for two weeks. They were less lucky at Mentone, but, as Tansley told his mother, the couple,

> managed to spend two delightful afternoons at La Mortola, where Sir Thomas Hanbury[31] has a most beautiful garden in which he grows an enormous number not only of mediterranean but of tropical and subtropical plants. He was very decent to us and took us round a considerable part of the garden.
>
> *Arthur to Amelia Tansley, 10 February 1904.*

From there the honeymooners visited Pisa, where 'we stayed in the same hotel that Bertie Russell and I stopped at 12 years ago', and then to Florence where 'E likes the pictures very much'. Rome, Naples and Sicily were next on their itinerary.[32]

***

Fritz Blackman was the best man at Tansley's wedding. Five years older than Tansley, he graduated in 1891, a year after Tansley went up to Cambridge. He was immediately appointed a demonstrator in the Botany School, and there he remained through a career in which he distinguished himself as a plant physiologist – he was one of the few foreigners to be made a Fellow of the American Society of Plant Physiologists.[33] While he claimed 'ecology was too difficult for *him*',[34] he willingly provided his brother-in-law with physiological expertise whenever it was needed. The two collaborated on many ventures, becoming affectionately known to Cambridge undergraduates as 'Black and Tan'.[35]

Living happily in St John's College and apparently destined for a life of bachelorhood, Fritz, at the advanced age of 51, surprised even those who knew him best by marrying Elsie Chick, herself aged 35, in 1917. Fritz found in Elsie a women who went some way in matching his own wide range of interests. Another graduate of UCL, but this time in English (BA in 1912, MA in 1914), Elsie had been President of the Women's Union, studied Icelandic, published academic papers, and in 1918 was (together with her sister Harriette) made a Fellow of UCL. Fritz was a man of catholic interests. A keen rugby football player when he was a young man, he was Steward of St John's, loved music, and was a keen collector of pictures. He was for many years a member of the syndicate that governed the Fitzwilliam Museum in Cambridge. Sydney Cockerell, the museum's Director, wrote:

> I can think of no member of the University to whom I was so much indebted for unflagging support and encouragement … I seldom drafted a report or fly-sheet without consulting him.[36]

Charlotte Mew's death in 1928 was a blow to the Chicks, Olivers, Tansleys, and Blackmans. It was Fritz who broke the sad news to Sydney Cockerell who, with his connections in the literary world, had been instrumental in helping Mew's poetry receive the critical acclaim it deserved. Sadly, recognition did not lift her out of the poverty into which she had fallen, or heal her despair over failures in love. Shortly after the death of her beloved brother, Mew killed herself by drinking lysol. Tansley is said to have been 'half-fascinated half-distressed by

Charlotte',[37] her unfolding life providing a first-hand case study for his awakening interest in psychoanalysis.

***

While a student at Cambridge, Tansley had been 'coached' by Sydney J. Hickson, Fellow of Downing College and author of *A Naturalist in North Celebes*. A 'nice' and 'very able' man,[38] he inspired Tansley with tales of his adventures, and the discoveries he had made, during his journey through the Malay Archipelago in 1885. Tansley was to see Malay for himself in 1900–1901. Up to this point, his botany had looked backwards, at the evolutionary history of plants and, in particular, at the origins of their vascular systems.[39] Through painstaking laboratory studies of plant tissues, involving many hours spent at his microscope, Tansley had made himself an accomplished anatomist. Reaching his 30th birthday only three months after his return from the tropics, he had come to a turning point in his career, as described in the next chapter. Henceforward, the focus of his attention would be whole plants, growing in the wild. Marvelling at the rich diversity of the plant life he had seen, and encouraged by Frank Oliver to believe that field botany in Britain could be more challenging and rewarding than laboratory studies of plant anatomy, Tansley was won over to ecology. He was about to go into the field.

Notes

1. Tansley Archives, Cambridge University Library, curriculum vitae.
2. Porter 1968, p.38.
3. Ibid., p.40.
4. Branscoll (a collection of family letters found at Branscombe, South Devon), 26 August 1897.
5. Porter 1968, p.38.
6. Ibid., p.48; Briggs 1948, p.651.
7. A few weeks before his marriage to Alys Pearsall-Smith, in December 1894, Russell told Alys that Tansley was the first of his friends 'whom I told about thee' (probably referring to their engagement and forthcoming wedding) (Griffin 1992, p.159).
8. Ralph Vaughan-Williams read history and music at Trinity College 1892–1895; much of his music celebrated the English countryside and its traditions.
9. Branscoll.
10. Ibid.
11. Tanner 1917, p.280.
12. Godwin 1977, p.4.
13. Hale Bellott 1929, p.373.
14. Tomlinson 1983, p.59.

15. Ibid., p.4.
16. Creese 1998, p.40.
17. Tomlinson 1983, pp.64–65.
18. Fitzgerald 1984, p.66.
19. Ibid., p.25.
20. Ibid., p.74; Andrew Roberts and Betty Falkenberg, 2005. 'Charlotte Mew chronology' available at http://studymore.org.uk/ymew.htm, archived under 'Mental Health History Timeline' in the UK Web Archive at http://www.webarchive.org.uk/ukwa/target/145406.
21. Tomlinson 1983, p.67; although for several decades women had been allowed to take courses and sit examinations, Oxford admitted women to full membership of the university, i.e. awarded them honours degrees, only in 1920. Cambridge resisted such a change until 1947.
22. Taylor 1968, p.38.
23. Ibid., p.37.
24. Ibid., p.36.
25. Ibid., p.39.
26. Godwin 1977, p.3.
27. Branscoll.
28. A little known botanist publishing, like Worsdell, on the comparative anatomy of grasses.
29. Branscoll.
30. Graduating from UCL in botany and geology, Stopes' first graduate research was with Frank Oliver on fossil plants (pteridosperms) found in coal balls. She later obtained a PhD (not the DSc mentioned in Harriette Chick's diary) with Professor Karl Goebel in Munich (Falcon-Lang 2008, p.133). After a disastrous marriage, she turned her attention to birth control, writing *Married Love* and opening her famous advisory clinics.
31. Mentone (= Menton) lies west of the French–Italian border, La Mortola just to the east. A rich silk merchant, Hanbury spent 40 years developing his celebrated gardens on a promontory overlooking the Mediterranean. In old age he bought a large garden at Wisley, Surrey, and donated it to the Royal Horticultural Society.
32. Branscoll, 10 February 1904.
33. Steward 1947, p.ii.
34. Godwin 1977, p.12.
35. Godwin 1957, p.232; 'Black and Tans' were World War I veterans recruited in 1921–1922 by the Royal Irish Constabulary to suppress – often violently – the movement for Irish independence.
36. Briggs 1948, p.653.
37. Fitzgerald 1984, p.72.
38. Branscoll, A. G. Tansley to his mother, 23 October 1891.
39. Tansley 1896a; Tansley 1896b.

6 Seashores and Woodlands: Looking for Patterns

> Plants are gregarious beings … they produce vegetation, as plant growth in the mass is conveniently called, which is actually differentiated into distinguishable units or plant communities.
>
> Arthur Tansley, *The British Islands and their Vegetation*[1]

Tansley often expressed regret that in his student days he had not spent time in the German laboratories, as had so many of the bright young botanists of his day. But he more than made up for this gap in experience when, between September 1900 and May 1901, he visited Ceylon, the Malay Peninsula, and Egypt. An immediate and tangible outcome was the small collection of ferns he brought back to University College, London (UCL), material whose anatomy was the subject of some of his earliest publications. One such investigation, 'On the structure of *Schizaea malacanna*' (1903), was, not insignificantly, conducted jointly with Edith Chick – the future Mrs Tansley. For the longer term, the species and range of vegetation that Tansley saw gave him a new perspective on plant communities. His experiences during those nine months were much more valuable to his future career than any time he might have spent in a German laboratory.

It appears that Frank Oliver was instrumental in providing Tansley with the opportunity to travel. One of Oliver's closest friends was William Lang.[2] They had become acquainted when Lang was working at the Jodrell Laboratory at Kew (close to Oliver's home), and they shared a research interest in fossil ferns. In 1899, Lang was a junior assistant in the Botany Department at Glasgow University. With the encouragement of his professor, F. O. Bower,

Shaping Ecology: The Life of Arthur Tansley, First Edition. Peter Ayres.
© 2012 by John Wiley & Sons, Ltd. Published 2012 by John Wiley & Sons, Ltd.

Lang set off that year on an expedition to Ceylon (Sri Lanka) and the Malay Peninsula with the intention of studying its cryptogamic vegetation, returning only in 1902. For nine months of that period, Lang was accompanied by Tansley. The inference is that Oliver, knowing of Lang's plans, prompted, or somehow persuaded, him to build Tansley into his plans. (Oliver would also have had to grant Tansley, his assistant at UCL, extended leave of absence.)

Until it was recently taken into the main library of the University of Cambridge, a large cardboard box labelled 'Tansley notebooks' sat on a dusty shelf in the small library of the Plant Life Sciences Department. It contained over 40 small pocket notebooks or diaries, each of which is crammed with Tansley's records of field sites he had visited, ranging in time from 'Ecological notes Norfolk Broads July 1902' to 'Cotswolds, 1929, 1934, 1939', and in location from 'Poland Czechoslovakia 1908' to 'Belgium 1933'. In pencil, or occasionally ink, he wrote neat, detailed notes of the plants he saw, the patterns of vegetation he recognised, and, often with the help of thumbnail sketches, the soils and topography he found. The earliest booklet is his 'Diary kept in the East 1900–01', started at the suggestion of Edith so that his letters home would not be short of anecdote or local colour.

In spite of its light-hearted conception, the diary reveals its writer was an obsessive keeper of records – of sea conditions during his voyages, temperature and relative humidity in his cabin, and seating plans at the captain's meal table. Surprisingly, given Edith's involvement, it reveals also that its writer took a special interest in which lady he was placed next to at the captain's dinner table, her dress, and the quality of her conversation. In mitigation, sea voyages were long and each passenger sought their own diversion from boredom. One particular voyage, between Penang and Colombo, was enlivened for Tansley because:

> the doctor on board was an old student of mine in London. It turned out to be Cadvan-Jones, who had recognised me as I was coming aboard in the sanpan; although I did not remember him at all. Jones was extremely affable during the voyage lent me books and papers, and we had several talks about U.C.L. ... There was a great dearth of musical talent so that I had to sing 'Tommy Atkins'[3] one night.

Tansley's enthusiasm for the main purpose of the trip is obvious from the very first entry in his diary. On Sunday 30 September he disembarked at 10:00 am from the SS *Oceana*, moored in Colombo harbour. His diary records that by mid-afternoon he was already busy collecting plants, such as *Ipomaea biloba* (= *I. pes-caprea*, goat's foot, Convolvulaceae), from the sandy shoreline a little south of the city. His entry for the same day notes, 'Saw first banyan tree exactly like pictures in children's books'.[4] The next day he took the train into the hills, to what was to be his base in Ceylon, the Royal Botanic Gardens at Peradeniya, near Kandy. There he met J. C. Willis, the Director of the Gardens (a Kew appointment), was shown the laboratory and herbarium, and was given help identifying the strand plants he had gathered.

The area around Peradeniya was the next to be investigated. By Sunday 7 October, he had met with Lang and they were exploring the dry jungle on a steep slope down to the Mahawli River. Tansley's diary reads:

> The trees are not tall – about 30–40 feet, and are rather sparse so that there is no difficulty getting through them except for the thin lianes [climbing plants] in which one is constantly getting caught. The first character that strikes one about the trees is the extraordinarily uniform type of leaf. This is rather coriaceous [leathery], and ovate, often narrowed at the apex with entire or very slightly toothed margin. Apparently this is characteristic of the leaves of all tropical jungles, not only the dry ones.

Learning all the time, the pair were soon travelling over the greater part of the island, sometimes by coach but often, particularly in Tansley's case, by bicycle. Willis was vital to their success because he could not only identify Ceylonese plants and supply background details of their distribution and habit, but he had invaluable local contacts. He could both advise Tansley and Lang where to go, and arrange transport and hospitality for them. Near Tangalla, on the south coast, the roadside hedges of high scrub, mostly *Lantana* and *Missenda*, were covered with abundant creepers, their appearance reminding Tansley of the *Clematis* that scrambled over the hedgerows of his beloved South Downs. He complained about the frequent early morning starts, at 5:00 or 6:30 am. By Saturday 3 November, he was so tired that, 'I stopped in bed, and afterwards watched a native cricket match on the green'. The hard work was interspersed with a smattering of dinners and garden parties, their frequency increasing in the later part of his travels when, without Lang, he travelled north through Egypt, from Suez to Cairo, his final destination before sailing for Marseilles, where he took a train to Paris and London.

In spite of social diversions, enough serious notes were gathered from Ceylon alone to justify two papers in the *New Phytologist* entitled, 'Sketches of vegetation at home and abroad. The flora of the Ceylon littoral' (1905). With the co-authorship of a UCL colleague, Felix Fritsch, who had special expertise in aquatic plants, Tansley set out to show that the coastal vegetation of Ceylon closely resembled that of the Indo-Malayan coast described by Schimper in *Die Indo-Malayische Strandflora* (1891).[5] In doing so he aimed to demonstrate the point that,

The flora of the hundreds of thousands of miles of coast bordering the tropical seas of the world is marked by a very striking uniformity of character and composition ... The botanist suddenly transported from the coast of Ceylon, for instance, to the shores of some island in the Malay Archipelago, two thousand miles away, could not tell that he had been carried further than round the next headland.[6]

Proving his point, there were clearly recognisable on the southwest coast of Ceylon four 'formations' (see Box 6.1) into which, Schimper had said, coastal associations of the Eastern tropics naturally group themselves: the *Pes-caprae* and *Barringtonia*, both on sand above the tide marks, and the mangrove and nipa palm formations, both on tidal mud. Tansley and Fritsch described minor differences between the coastal vegetation of Ceylon and that of Indo-Malaya but Schimper's categories held good in both areas.

While the death of Queen Victoria on the 22 January 1901 and the dawning of the new Edwardian age was occupying everyone's thoughts in Britain, far away in the East Arthur Tansley was awakening to a new age of his own. Every day's exploration confirmed not simply the great variety and richness of tropical vegetation, but his growing feeling that patterns could be recognised and order brought to the description of vegetation – and if that were possible for luxuriant tropical vegetation, it must be equally possible for vegetation much closer to home.

***

Frank Oliver, the key figure who encouraged Tansley to relinquish plant anatomy in favour of ecology, was by all accounts an exceptionally nice man. Born into a sober Quaker family, Oliver was a modest, unassuming man whose achievements have tended to be overlooked. When the history of botany has been written, Oliver has been given only a supporting role. Characteristically, it was Oliver who assembled and edited a celebratory volume, *Makers of British Botany* (1913), that included several of his contemporaries in its roll of honour. Oliver has a strong claim to be included among those 'Makers', for while there is no single discovery or concept attributable to him, he was central to the emergence of ecology, inducting a generation of young men and women into its ways. His interests would not be out of place today for they included the botanical effects of air pollution and the invasion of British habitats by alien species. Seeing clearly an applied role for botanical science, he was a tireless worker for the conservation movement, with a record of successful outcomes (Chapter 9).

After a childhood spent in and around Kew, Oliver studied botany at Cambridge in the mid-1880 s, a period when unfortunately for him the teaching staff was, with the notable exceptions of Sydney Vines and Francis Darwin, uninspiring. What Cambridge failed to supply he found during two long vacations spent studying in Germany. In the summer of 1885 he was at the

Bonn laboratory of the cytologist Eduard Strasburger, and in 1886 at Tübingen where, under the tutelage of the physiologist Wilhelm Pfeffer, who was investigating the causes of plant movements, he was guided to study the contractile stigmas of *Mimulus glutinosus*. It was an interest that he continued to pursue after his return to Kew, where he recognised and researched the contractile labellum of a Colombian orchid, *Masdevallia muscosa*. But there was another more important legacy from his German summers: in Bonn he made friends with the plant geographer, Andreas Schimper (Chapter 1), who kindled his interest in ecology. Schimper, still less than 30 years old at the time of their meeting, had already travelled widely in the tropics of South America. While briefly holding a Fellowship at Johns Hopkins University in Baltimore he had also explored the temperate lands of North America.[7] More than ten years were to pass before Schimper would publish his seminal *Pflanzen-geographie* but his ideas were falling into place and in Bonn he was able to impress indelibly on Oliver's mind the fundamental principle that ecological explanations must have a physiological basis.

Oliver's ecology was, like Schimper's, slow to bear fruit. Appointed to a lectureship at UCL in 1888, there followed a lengthy period of about 15 years in which he spread his talents widely. Thus, for several years it was plant anatomy, particularly as it related to developmental series of fossil plants, that was his primary concern. In collaboration with D. H. Scott, Honorary Keeper of the Jodrell Laboratory at Kew, he studied the reproductive organs of fern-like fossils (Pteridosperms) from coal measures of the Carboniferous period.[8] It was with such anatomical studies that he involved Edith Chick, Marie Stopes and, for a while, Tansley. On behalf of the Royal Horticultural Society, Oliver carried out a detailed investigation of the effects of urban smog on cultivated plants, a study which allowed him to employ his knowledge of plant physiology. Although concerned with glasshouse crops, the study had many implications for applied ecology for this was a period when smoke-laden fogs, which enveloped London and other major cities for weeks on end every winter, regularly damaged the leaves of evergreens and killed the resting buds of trees and shrubs.[9] (Conditions were so bad during the worst fogs that extra gas lights had to be placed strategically in the larger lecture theatres at UCL.[10]) Oliver set up pollution monitoring sites around the capital[11] but could suggest nothing to protect the plants of streets or gardens. Anticipating measures widely taken many decades later, however, he recommended that horticulturalists should grow their plants in air-tight glasshouses with air fed in through a charcoal filter to remove pollutant gases.[12]

On clement days Oliver brought a touch of fresh air to the lecture room, quite literally. Rapidly dismounting from his bicycle after arriving late and breathless, he would fling open the lecture room window and address the class from that position.[13] Metaphorically, too, he seems to have been a breath of fresh air, attractive to both students and younger colleagues. With a tongue-in-cheek sense of humour, he remained something of a rebel throughout his life; 'disliking all authority, rules and regulations ... [he] liked to be a law unto himself'.[14] An energetic mountain climber and walker, many of his vacations were spent in the Channel Islands or on the nearby French coast where he was quickly impressed by the mobility of the dune and shingle systems. From about the turn of the century, Oliver devoted more and more of his working time to the study of such dynamic systems, all the time involving Tansley. Indeed, he may have been carried along by his young assistant's enthusiasm.

Tansley's earliest dated field notes after his return from the tropics were 'Ecological notes on the Norfolk Broads, July 1902'. It is not known whether his visit was planned as the precursor to a student excursion, but in July 1903 Oliver led a small party of advanced students and staff from UCL to the eastern Norfolk Broads, where the party lived and worked on two sailing boats.[15] In the summer of 1904 he led a larger mixed party of students from UCL on a two-week field trip to northern France. If not familiar from his own recent ramblings, the area may have been known to him since childhood for his father, Daniel, took many vacations in France where the light was ideal for his hobby of painting.[16] The site selected for study was an estuarine one, at the Bouche d'Erquy, in northern Brittany.

> A certain Professor of Botany
> To save his class from monotony
> Lead his students a dance
> Round a salt marsh in France
> To develop their brains
> If they've got any.
> *An anonymous student*[17]

Oliver's proposal to hold a lengthy field trip for students was significant because it recognised that, in words written by Tansley many years later, 'ecology is essentially a *practical* study, the benefits of which must first be sought in the field, not in theoretical instruction in the classroom'.[18] Lengthy field trips for students were rare but not unprecedented in 1904. As already seen, Tansley had enjoyed such an expedition with the Cambridge geologist, McKenny Hughes, when, for two weeks, each day was spent in the field and each evening was spent examining and discussing the rocks and fossils that had been collected.[19] In the USA, a precedent for an *ecological* excursion had been set when Henry Chandler Cowles originated such a course, 'Botany 36 (Field Ecology)', at the University of Chicago in 1900–1901. Beginning on a small scale with 'Ecological Anatomy and Field Botany' in 1897–1898, Cowles had developed a complete ecological curriculum;[20] with 'Botany 36' he would each year take a party of students for four weeks to an area of ecological interest. These were often around the Great Lakes, where he could pursue his own research.[21] Cowles was (like Tansley) an admirer of both Schimper and Warming.[22] The latter had made a minute study of the glaciated Danish landscape, where drifting dunes were being anchored by vegetation, and lakes were converted to swamps and bogs, systems that were patently dynamic and the principles of which Cowles easily recognised in his own Great Lakes area.

When students signed up for Cowles' 'Botany 36' they received a set of instructions with a warning, 'It is imperative that each member of the party be a "good sport", putting up cheerfully with rain, hot or cold weather, mosquitoes, black flies, and with inadequate or unsatisfactory accommodations'.[23] Students must have 'stout tramping shoes, with leggings, if the shoes are low' and both sexes needed trousers. The 'most satisfactory garb for women', Cowles wrote, '… is a riding habit made of khaki or other suitable material'. Possibly because conditions likely to be experienced on the north coast of France were less extreme than those of mid-west America, the UCL women dressed more demurely, though less practically, than their American sisters. Photographs of Oliver's Bouch d'Erquy expeditions of 1904 and 1905 show them wearing long dresses, and the large hats so fashionable in the Edwardian era, while conducting research on the salt marshes (Figure 6.1).

Figure 6.1   The field laboratory on the occasion of the first visit to Bouche d'Erquy, Brittany, 1904. Soil weighing and salt determinations are in progress in the foreground. Others figures are boring for soil cores. From left to right: E. A. N. Arber, W. C. Worsdell, F. F. Blackman, and 'at the balance room' Frances Chick. (From *New Phytologist* 1906, with permission.)

The Bouche d'Erquy was chosen by Oliver because access to it was easy – which had not always been the case the previous year in Norfolk. Also the Bouche was relatively flat, and there was a very sharply defined characteristic flora consisting of comparatively few species whose distribution, it appeared on first inspection, was largely dependent upon physical factors that might be measured. The estuary was separated from the sea by a spit of old sand dunes, extending at right angles to the main direction of the river and allowing its exit to the sea via a relatively narrow channel, which itself was fed by three smaller streams. On higher ground, some distance from the sea, there were five dominant salt-tolerant species, *Salicornia herbacea*, *Glyceria maritima*, *Suaeda fruticosa*, *Obione portulacoides*, and *Juncus maritimus*. This simplified the primary work of the party which was, in 1904, to map in three dimensions the physical features and the plant associations. Simultaneously, preliminary measurements were to be made of physical and chemical characteristics of the soils.

The 'survey', for significantly that is what it was called, was very carefully based on the 'method of squares'. A convenient baseline was chosen, as in an ordinary land survey, and squares with 100-foot sides were marked on either side. Each square was carefully marked with labelled flags and then subdivided by sticks placed at 20-foot intervals. If the grid was constructed with perfect right angles, the flags lined up along diagonals as well as along each side. Data from each square were transferred onto a map drawn on a scale of 1:240. Squares of the greatest interest were themselves subdivided into a 'gridiron' of smaller squares, each with sides of five feet. Data from smaller squares were mapped on a larger scale of 1:60. (Examples of the resultant maps are seen in figures 78 and 80 of Oliver and Tansley (1904).) The bulk of the party was divided into five working sections, each of three members who acted as 'surveyor', 'diarist', or 'collector'. Each section dealt with a different part of the site. The remainder of the party made the physical and chemical measurements of the site, photographed characteristic plants and associations and, in the case of two specialist artists, pictured in water colours the salt marsh landscapes.[24] Oliver and Tansley summarised their approach in 'Methods for surveying vegetation on a large scale', published in *New Phytologist* in December 1904.

With the good humour of youth, Tansley recorded that field work had been disrupted on one occasion by 'officious French visitors' to the area who 'conceived it to be their duty to interfere and protest against the evidently nefarious designs of the party upon the well-being of the Republic, cunningly concealed by the pretext of botanical investigation'. The situation was defused only when local officials explained the bona fide purposes of the Britons. The other problem the party encountered was that their flagged bamboo markers proved of irresistible fascination to the 'local peasantry and fisher-folk' with the result that they were having to be regularly replaced.[25]

During the winter of 1904–1905, the sheets of the map were reduced to 1:480 and a rough abstract was drawn at the even smaller scale of 1:2500. Staff and students could now begin to draw conclusions. For one, there was a very clear correlation between the vegetation and height above sea level, the latter determining the water and salt content of the soil.

The following Easter in 1905 the site was revisited and all important physical changes were mapped. In selected one-foot squares, seedlings were counted. They were counted again at Whitsun and once more when the main party from UCL returned in September.[26] The plan for 1905 was to pay more attention to the soil by making daily measurements of its water and salt content through complete tidal cycles. While 'Mr Tansley' was charged with charting the vegetation, with the help of five ladies, 'Dr [F. F.] Blackman', with the help of Mr Baker and six ladies (including Miss Frances Chick), was charged with gathering the physical measurements. Oliver, assisted among others by Miss Stopes, Mr Hill, and Mr Salisbury, was to continue contouring the general map of the area. Evenings were to be spent in the exchange of information between groups. The expenses of the journey, out and home, and of the stay, were estimated by Oliver to be approximately £4 per person.[27]

The carefully planned programme for 1905 was successfully completed and parties of UCL students made further visits to Bouche d'Erquy in 1906, when an indoor laboratory and a store for apparatus were set up in a nearby cottage, and also in 1907. The laboratory enabled greater emphasis to be placed on physiological measurements, such as of the osmotic pressures of root cells. Year by year greater emphasis was placed on experimental work in the field as, for example, when seedlings were transplanted between zones to examine their survival. The effects of soil chemistry on seed viability were examined. Also, an experiment was made to determine whether *Spartina alterniflora*, which was already invading large areas of the mud of Southampton Water and the Solent, would colonise the mud flats of Erquy. The results were inconclusive because all traces of the introduced rhizomes of *Spartina* were lost. However, by the final year at Erquy other changes were evident: for example, at the northwest corner of the study area, vegetation had advanced about 40 feet on previously uncolonised sand, with *Salicornia radicans* and *Suaeda maritima* leading the way. Although partly a teaching exercise, the Erquy studies embodied new and important research whose importance was realised when Oliver was asked to advise the Royal Commission on Coast Erosion and Land Reclamation about the part played by halophytes and sand- and mud-binding plants in combating erosion. It was one of the earliest recorded instances of an ecologist being called as an 'expert witness' by a government body.[28]

By the second year of the Erquy expeditions, 1905, significant developments were afoot in the wider world of ecology. At a practical level, Frederic Clements, who had learned his ecology under the supervision of Charles Bessey (Chapter 1), published in that year his *Research Methods in Ecology*. It was based on Clements' own studies of the grasslands of Nebraska (his home state) and of Colorado. One of his basic sampling tools was the quadrat, a one-metre square frame used to delineate a patch of vegetation and soil to be studied in detail. Acknowledging Clements' lead, Oliver's 'General Scheme of Work' for 1905 proposed that permanent 'quadrat-stations' would be laid out at Erquy. To this day the quadrat remains fundamental to the practice of botanical map-making.

When Tansley addressed the British Association for the Advancement of Science (BAAS) meeting in 1904 in Cambridge, he laid out what he believed were 'The problems of ecology'.[29] A vision of the future of ecology emerges from that address, and also from a review of Clements', *Research Methods in Ecology*, that he and F. F. Blackman wrote for the *New Phytologist*.[30] The vision distinguished two stages. The first, being a descriptive stage – based on survey – in which plant associations are 'characterised, enumerated and described'. The second, philosophically more demanding and therefore potentially more rewarding, being when the causes of phenomena are unravelled.[31]

Clements thanked Tansley and Blackman for 'your fair, critical and yet appreciative review in the "New Phytologist" ... I am especially grateful for the opportunity of again testing my views in the light of your criticisms'. His letter was dated 12 January 1906, but he and Tansley had been corresponding for at least the previous six months for, on 30 July 1905, Clements had been telling Tansley how he was worried that most of his fellow Americans were still in the stage of 'descriptive ecology'.[32] Clements was more judgemental than Tansley. He labelled the first of Tansley's two stages as 'floristic', a derogatory term. It was all too often the territory of the dilettante: 'The bane of recent development popularly known as ecology has been a widespread feeling that anyone can do ecological work, regardless of preparation'.[33] The second stage was true ecology. 'The organic connection between ecology and floristic has produced an erroneous impression as to the relative value of the two. Floristic has required little knowledge, and less preparation; it lends itself with insidious ease to chance journeys, or to vacation trips, the fruits of which are found in vague descriptive articles'.[34] Clements saw no boundary between ecology and physiology, apart from man's current state of ignorance, and believed that in the future the two would fully merge.

Tansley was more patient. While recognising that the second stage would involve observation and experiments by professionally trained men and women, and while looking forward to the day when there would be permanent laboratories established in regions exhibiting specialised types of vegetation,[35] he saw in his pragmatic way that it could not be entered upon immediately. The first phase had still to be completed and to do that, he thought, the community of amateur botanists might, with the right organisation, be usefully harnessed.

Field botany had become popular in Britain by the end of the 19th century. Participants ranged from gentile ladies who were encouraged to draw flowers, to artisans from the new industrial cities who, on their one day of the week free from labour, congregated in pubs before essaying into the countryside to study the flora – a pattern of behaviour christened 'botany at the bottom of a glass'.[36] The ways in which these amateur botanists were organised ranged from not-at-all to highly, examples of the latter being those who belonged to widely respected bodies such as the Yorkshire Naturalists' Union (which spawned the British Mycological Society in 1896) or the Woolhope Naturalists' Field Club based in the small cathedral city of Hereford. Botanists who aspired to professional standards could join the Linnean Society (founded in 1788) or the Botanical Societies of London or Edinburgh (both founded in 1836).

Tansley regretted that 'one of the most crying examples of the waste of good work and sound knowledge in the field of modern botany, is the utilisation of the work of the local botanist and the local field club. … Convince them of the interest of ecological survey work, and you would secure their co-operation in working out and mapping local floras … they would do a hundred times better than a visiting botanist, with no knowledge of the locality'.[37] 'For these reasons', he wrote in 'The problems of ecology', 'I would like to see a central committee formed for the systematic survey and mapping of the British Isles'. Late in the same year, 1904, the Central Committee for the Survey and Study of British Vegetation was founded when Tansley, C. E. Moss, and T. W. Woodhead met at William Smith's house in Leeds. Smith was appointed Secretary, and five invited but absent botanists were included in the inaugural committee. Excepting Lloyd Praeger from Dublin, all Committee members were college lecturers, although they had reached their positions by diverse and often indirect routes. Thus, Moss had taught at a school in Somerset and later at a teachers' training college in Manchester (while he enrolled for advanced courses at Owens College). Woodhead had worked for a woollen company in Huddersfield before becoming a lecturer at the local Mechanics Institute.[38]

Members of the Committee typically lacked a formal academic training, being largely self-taught, but each brought to the Committee considerable practical experience. Moreover, each was a teacher, someone who had had to reflect on, distil, and organise his knowledge. They would have regarded themselves as professionals. Protecting the status of their new Committee and diverging from Tansley's original idea, its members agreed it was not 'desirable to try to obtain workers … through natural history societies etc' but they would, instead, try 'to obtain suitable workers to co-operate with the Committee'.[39] Control should not be ceded to amateurs, or their clubs or societies. In the 19th century, professional taxonomists in Britain had faced a comparable problem: the need to maintain control of their subject – which in their case involved the naming and classification of the flood of new plants arriving from all corners of the Empire – while encouraging local collectors and expeditions sent out from Europe. Joseph Hooker had successfully protected the *science* of taxonomy, and established Kew as the highest authority in the subject, by laying down a strict, often ruthless, set of rules and rewards for collectors.[40]

The challenge facing Tansley was to ensure that the Committee carried out those surveys of British vegetation that were so urgently needed without losing sight of the higher objectives of ecology, or losing control of the *science* of ecology. He had already strengthened his position among professional ecologists when, in 1902, he had founded the *New Phytologist* journal, which he would edit for the next 30 years. Whether he had a clearly thought out long-term strategy, or was merely adept at recognising the right time to make each of a series of discrete moves, Tansley was successful. In 1911 he made connections with the international community of academic ecologists when he organised the first International Phytogeographic Excursion (IPE I), which was held in Britain. And in 1913, the British Ecological Society (BES), the first professional organisation of its kind in the world, was founded with Tansley as its President. With these moves he secured for ecology a respectable place within the mainstream of the biological sciences. But this is to anticipate the next chapter; the immediate task for the Central Committee (known from 1911 by the briefer but no more exciting title, the Vegetation Committee) was to forge working relationships among its members, and in particular for Tansley to harness the experience of William Smith who had already published maps of the vegetation of large parts of Yorkshire.

William Smith's education was, in contrast to Tansley's, provincial – he and his brother, Robert, had attended the University College, Dundee, where they studied with Patrick Geddes[41] – and William was obliged to work throughout his life. He held various teaching posts, in a school and universities, obtained a PhD in plant pathology in Munich, and in 1897 moved to Leeds where he was appointed Lecturer in Agriculture and Assistant Lecturer in Botany at the Yorkshire College, soon to become the University of Leeds.[42]

It is not certain whether Tansley or Smith convened that first meeting of the Central Committee at Smith's house, but from the outset their involvement seems to have been well balanced. Both sat on every one of the six sub-committees, whose various purposes were to decide matters such as how to obtain grants for botanical surveys and the appropriate scale, symbols, and colours that might best represent different types of vegetation on maps. Tansley found much to admire in Smith who combined 'the solidity and pertinacity of the best type of Lowland

Scot' with a modest nature, never seeking credit for himself and always willing to help others.[43] Their paths diverged in 1908 when Smith's primary interest turned to agriculture and he left the Committee.[44]

William had been introduced to survey work by his brother, Robert, who was another of Warming's many admirers and a keen follower of the Dane's methods. Robert's other hero was Charles Flahault, with whom he had been lucky enough to spend the winter of 1896–1897 at the University of Montpellier – an arrangement facilitated by Geddes.[45] On returning to Scotland, Robert began to plan how he might map the Pentland Hills, just south of Edinburgh. Following Flahault's example he would delineate 'natural regions' on the basis of their dominant species. The natural forests of southwest France, however, made it relatively easy for Flahault to pick out the dominant species (typically trees), whereas Robert had to deal with a countryside that was all too obviously the product of man's activities, necessitating a more common sense approach in which grasses and heather moors were among the main divisions. Tragedy struck in 1900 when Robert died of peritonitis. He was aged only 26, but he had completed his maps of the Edinburgh and of the North Perthshire regions and believed that from this small sample he was already able to distinguish littoral, temporal, sub-alpine, and alpine as the main types of vegetation in Scotland.[46]

Robert's unfinished surveys of the Forfar and Fife regions were completed by his brother William during vacations from the Yorkshire College. William must have been a busy man for, immediately on his arrival in Leeds in 1897, he had joined the Yorkshire Naturalists' Union, persuading its members to begin a Botanical Survey with himself as its convenor and Secretary. Further impetus was given to the Yorkshire surveys by the enthusiastic reception given to the paper describing Robert's work that his brother read to the 1901 meeting of the BAAS. This paper was described by Woodhead[47] as a breathe of fresh air that took botanists out of the laboratory into the field. When the Yorkshire vegetation maps appeared in 1903 they were the first for any county in England and covered 1700 square miles, or roughly a quarter, of its largest county. William had been greatly assisted in this huge undertaking by Charles E. Moss and William Munn Rankin, each with unsurpassed knowledge of their local area – Leeds and Halifax, and Harrogate and Skipton, respectively. Both were founder members of the Vegetation Committee.

William Smith's objectives as Secretary of the Committee were 'to coordinate the work being done, to secure uniformity of method so far as may seem desirable, and to have ready means of discussing various topics that arise in connexion with methods and results'.[48] Certainly with regard to the last objective he was successful for, when writing 'The early history of modern plant ecology in Britain' (1947a), Tansley recalled, 'I have never attended meetings at which the interest was keener and more sustained'.[49] This happy memory is followed shortly, however, by a more critical section in which he minimises the achievements of the Committee in its first five years:

> neither the primary surveys nor the more intensive studies, can be said to have been directly inspired by the meetings of the British Vegetation Committee. They were all independent efforts, in many cases undertaken before the Committee was formed. The committee certainly stimulated interest and enthusiasm in high degree and the members learned a great deal from one another; but it was not until 1910 that the direct results of its activity became apparent. ... In that year Moss published in the *New Phytologist* 'The fundamental units of vegetation'... and in the same year, and in the same journal, there appeared an account of 'The woodlands of England, natural and semi-natural' by Moss, Rankin and myself.[50]

Was Tansley too critical? What did happen in those years between the founding of the Committee in December 1904 and the *New Phytologist* publications of 1910? On the positive side, within its first 12 months the Committee issued a six-page pamphlet aimed at novices and entitled, 'Suggestions for beginning survey work on vegetation'. In November of the same year, 1905, members met in Liverpool to give short informal papers describing their current surveys and to display recently published maps. An obvious highlight of the meeting was the demonstration by Praeger and his colleague Pethybridge of their highly detailed, coloured maps of the area immediately to the south of Dublin, the first survey of any part of the wetter regions of western Britain. The Dublin pair argued that species were 'naturally aggregated into a number of vegetation-types or synthetic plant groups, which recur within the area wherever similar conditions of environment exist'.[51]

Another survey that was under way when the Committee was formed was that of the Scottish Highlands carried out by Marcel Hardy (an ex-student of Flahault). His coloured vegetation map was extraordinarily detailed (one inch to the mile or 1:63, 360) and displayed 12 types of vegetation. Each was distinguished by the dominant and sub-dominant plant species and, in associated publications, Hardy related each to geology, soils, and climate, and to local forestry and farming practices.

As the number of maps, plus 'memoirs' or accompanying set of notes, built up, members of the Vegetation Committee realised their endeavours were being held back by the high costs of colour printing. Noting that in producing his maps of the south of Dublin Praeger had received financial assistance from the Department of Agricultural and Technical Instruction for Ireland, the Committee asked Tansley to seek help from the Board of Agriculture, the body responsible for the Ordnance Survey mapping of England, Wales, and Scotland. Between 1907 and 1910, Tansley met and exchanged letters with officials of the Board of Agriculture, at one time being given indications that help would be forthcoming, but finally being told that neither the Ordnance Survey nor the Stationery Office could be responsible for issuing publications which, they judged, emanated from a private body with no official status. His quest to acquire help for mapping proved fruitless. It was one of the few endeavours in Tansley's life in which his efforts went unrewarded.

If there was no money for printing, and if most of the mapping undertaken was commenced before the Committee formed, was it a failure? The short answer is no. It was not a failure because, if nothing else, the Committee initiated IPE I (1911) and in 1913 metamorphosed into the first, and founding, Council of the British Ecological Society (Chapter 7). These major steps forward were facilitated by changes in the membership of the Committee, the interests of its newer members being more in sympathy with those implicit in the second stage of Tansley's vision for ecology, the one in which the causes of phenomena are unravelled.

An early step in that direction occurred in the first few months of the Committee's life when Oliver was recruited. He brought with him not only the experience that he and Tansley were accumulating through their Norfolk and Erquy surveys but, more importantly, ambitions that went far beyond simple mapping. Tansley had gained a natural ally on the Committee. (It is indicative of Oliver's modesty and Tansley's self-confidence that the junior man was happy to bring onto the Committee his direct superior at UCL; with other men such a combination might have proved a recipe for deference or wrangling, either of which would have inhibited the work of the Committee.) Marcel Hardy left in 1906 but the Committee was joined by F. E. Weiss and R. H. Yapp, who were already studying the physical environment and vegetation of Wicken Fen. When William Smith and Pethybridge, another founder member, left the Committee their replacements were similarly much less involved with large-scale survey work and much more involved in studies of the relationship between the environment and local vegetation.[52]

The one founding member of the Committee who, more than any other, shared Tansley's interests in causes was Thomas Woodhead. Deemed by Tansley to be a pioneer in the ecological study of vegetation in Britain,[53] Woodhead had by 1904 already made a detailed study of the relationship between the soil profiles and vegetation in Birks Wood, Yorkshire. His base was the biological laboratory of the nearby technical college in Huddersfield. However, he erected huts within the woodland allowing him to conduct some research on site and, possibly, setting the example for studies soon to be made by Oliver and Tansley in Brittany.

Birks Wood lies on a northwestern-facing slope, its soils showing a gradual transition from a shallow, well-drained sandy loam on the northwestern edge to a heavy clay loam, with at least 150 centimetres of surface humus, at the higher southeastern edge. Originally an oak wood, some elm, sycamore, and beech had invaded, as too had bracken. Wood soft grass (*Holcus mollis*) was also widespread. The frequency of bluebell, upon which Woodhead first focused his attention, increased along a north–west to south–east gradient. Such a detailed study of woodland vegetation, based on topography and soil type, was in itself ground-breaking but, in addition, Woodhead was able to show for the first time a relationship that he called 'complementary association'. In areas where bracken, wood grass, and bluebell were all common, that is away from the edges of the wood, the species minimised their competition in two ways. First, they grew and flowered at different times of the year: bluebell in spring, wood grass in early summer, and bracken in late summer. Second, they occupied different depths in the soil: bluebell bulbs were deepest, bracken rhizomes at an intermediate depth, and wood grass roots close to the surface.[54]

In 1907 Tansley moved from UCL to a lectureship in Cambridge. He was able to collaborate there with another Committee member, Charles Moss, who had recently been appointed Curator of the Herbarium in the Botany School.[55] With Tansley's help, and on behalf of the Committee, Moss wrote a paper 'The fundamental units of vegetation', which was published in Tansley's *New Phytologist* (1910). It was a bold attempt to address a problematic area that had continually worried the Committee. Moss proposed three hierarchical units of vegetation (Box 6.1). 'Plant formation' should be used to describe the whole of the vegetation occurring in a well-defined, uniform habitat, while 'plant association' should be used for areas of variation within the formation. The smallest unit would be the 'plant society'. The three could be neatly represented in quarter inch (1:253, 440), one inch (1:63, 360) and six inch (1:10, 560) to the mile maps, respectively. Unfortunately, when the proposals were presented to the International Congress of Botanists at Brussels (1910), Moss and Tansley found many objectors. Among the leading European critics were Flahault and Schröter who believed the proposals did not translate easily to the more diverse conditions found in continental Europe, and suggested that the influence of climate on soils and vegetation was not given sufficient recognition. Their objections were, perhaps, not surprising because long ago Schimper had conditioned European thinking when he proposed that the largest units of vegetation were determined by climatic factors (climatic formations), the vegetation so determined being separable into smaller units (edaphic formations) by soil type. Moss had simply argued that the British climate was so uniform as to justify the primary classification of vegetation according to soil type; these units were his 'plant formations' and in Tansley's opinion equivalent to Schimper's 'edaphic formations'. It was a disagreement which Tansley had to face again the next year when European and, in addition, North American botanists visited Britain to take part in IPE I (Chapter 7).

Box 6.1   Plant communities

Formations

Formations are composed of plants with a similar body or 'life form', wherever they occur. Examples of life forms are the large deciduous tree, the small sclerophyllous shrub, the meadow grass, and the tall slender swamp plant.

A formation type will differ in detail according to location. Thus, the equatorial rain forest formation differs between Indo-Malaya, Africa, and America because at least some of the component species will be different.

Associations

Within a formation there are a number of associations, smaller communities recognised according to the species by which each is dominated.

Societies

Societies are small communities dominated by species not dominant in the wider community. Thus, within oakwoods, ash or alder may dominate locally on wetter soils.

Succession

Vegetation is dynamic; the first plant communities are gradually suppressed and superseded by others so that a series or succession results, culminating in the climax community where the vegetation is in equilibrium with the whole of the environment, for example the equatorial rain forest.

Sere

The series of communities leading to a climax is a sere; the complete, uninterrupted, 'natural' succession from bare habitat to climax is a primary sere, or prisere.

Box 6.1 is based on chapter 10 of Tansley's *The British Islands and their Vegetation* (1939a), though as that text shows the ideas had been discussed for decades. The concept of succession was first clearly set out in the work of the North American ecologists H. C. Cowles (1899)[56] and F. E. Clements (1916, 1928),[57,58] a provenance Tansley always acknowledged, though he developed arguments against Clements' idea of a single (climatic) climax. He proposed instead that, at least for the British islands, edaphic and biotic climaxes were equally important – as will be seen in Chapters 7 and 9.

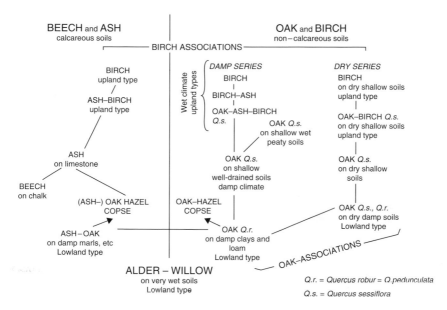

Figure 6.2   Relations between the dominant species of tree and conditions determining the associations. Upland types are at the top, lowland at the bottom. The size of the capitals indicates the importance of the association. (After Moss, Rankin, and Tansley 1910.)

Meanwhile, determined to press forward but bearing in mind the hitherto largely overlooked effects of man, Moss, Rankin, and Tansley published in *New Phytologist* (1910) a lengthy paper, 'The woodlands of England'. Three years earlier, Tansley had drawn on the results of studies of woodlands gathered by members of the Vegetation Committee to suggest a unifying theory. He proposed that there was a primary division between woods of *Quercus sessiliflora* [sic] (= *Q. petraea*, sessile oak) and those of *Q. pedunculata* (= *Q. robur*, pedunculate or common oak) which relied upon differences in soil moisture, the former being found in regions with drier soils, the latter in regions with moister soils. The three authors now extended this scheme to include most of the common types of woodland. The dominant type depended, thus, on altitude and the east–west transition in climate. As illustrated in Figure 6.2, wetter soils favoured an alder–willow series, with the lime content of the soil proving a modifying influence. On drier soils, an oak–birch series was found on siliceous soils, while a beech–ash series was more common on lime-rich soils.[59] The scheme took full account of human interference, such as the effects of coppicing, and was one of which Tansley was justifiably proud. He wrote in 1947, 'even today, there is very little to alter, though there is much to add'.[60] Woodland ecology was in 1910 probably the most advanced branch of ecology, thanks in large part to Moss, Rankin, and Tansley who had laid its foundations.

Patterns were emerging. Now it was time for Tansley and the Committee to share with the international community what they were learning about the vegetation of the British islands.

Notes

1. Tansley 1939a, p.213.
2. William Henry Lang was to find fame when, with his Glasgow colleague, Robert Kidston, he described the anatomy of primitive plants fossilised in the Rhynie Chert deposits near the village of Rhynie, Aberdeenshire. Their publications in 1917–1921 went far to explain the origin of land plants. In 1909 he was made Professor of Cryptogamic Botany at the University of Manchester (Salisbury EJ. 1961. William Henry Lang 1874–1960. *Biographical Memoirs of Fellows of the Royal Society* 7, 147–160).
3. A popular music hall song of the 1890 s about the common British soldier; its origins are obscure but may lie as far back as the 18th century.
4. 23–27 February 1901. *Diary kept in the East.* A18 Tansley Archives, University of Cambridge Library.
5. Schimper AFW. 1891. *Die Indo-Malayische Strandflora.* Jena: G. Fischer.
6. Tansley, Fritsch 1905, p.1.
7. Salisbury 1952, p.230.
8. Ibid., p.232.
9. Oliver FW. 1891. The effects of urban fog upon cultivated plants. *Journal of the Royal Horticultural Society,* **13**, 139–151.
10. Lester 1995, p.68.
11. The apparatus employed had been developed by G. H. Bailey at Owens College for use in Manchester (J. Sheail, personal communication).
12. Oliver FW. 1893. The effects of urban fog upon cultivated plants. *Journal of the Royal Horticultural Society,* **16**, 1–59.
13. Salisbury 1952, p.231.
14. Ibid., p.230.
15. Tansley 1903, p.167.
16. Oliver 1927, p.9.
17. Salisbury 1952, p.234.
18. Tansley, Price Evans 1946, preface.
19. Godwin 1957, p.229.
20. Cassidy 2007, p.41.
21. Ibid., p.80.
22. Keen to know the details of Warming's writing, and not wanting to wait for a translation of *Plantesamfund,* Cowles taught himself Danish (Ibid., p.28).
23. Ibid., p.82.
24. Tansley 1904b, p.204.
25. Ibid., p.204.

26. Oliver FW. 1905 (unpublished). *Erquy Expedition. An open letter dated 28 June 1905.* Tansley Archives, University of Cambridge Library.
27. Anon. (unpublished). *Botanical Expedition to the Bouche d'Erquy. General Scheme of Work.* Tansley Archives, University of Cambridge Library.
28. Oliver 1906; Hill 1909.
29. Published as Tansley 1904a.
30. Blackman, Tansley 1905.
31. Tansley 1904a, pp.195–6.
32. H26 Tansley Archives, University of Cambridge Library.
33. Ibid.
34. F. E. Clements, cited by Blackman, Tansley 1905, p.220.
35. Tansley 1904a, p.200.
36. Secord 1996, p.381.
37. Tansley 1904a, p.198.
38. Sheail 1987, p.27.
39. Ibid., p.23.
40. Endersby 2008: *inter alia*, Hooker rewarded with the gift of a microscope those colonial botanists who collected for him interesting specimens of a high standard (p.75).
41. In 1885, while lecturing in zoology at Edinburgh University, Patrick Geddes offered summer schools in botany and zoology at the Granton Marine Station, near Edinburgh. These were not, however, an integrated part of a university course. In 1888 he became Professor of Botany at University College Dundee; see also note 40, Chapter 12.
42. Sheail 1987, p.6.
43. Ibid., p.22; Tansley 1929a, pp.172–173.
44. Schulte 2000, p.308.
45. Sheail 1987, p.8.
46. Tansley 1947a, p.131.
47. Sheail 1987, p.11.
48. Tansley 1905, p.28.
49. Tansley 1947a, p.133.
50. Ibid., p.134.
51. Sheail 1987, p.24.
52. Tansley 1947a, p.133; Schulte 2000, p.308.
53. Tansley 1905, p.26.
54. Sheail 1987, fig. 1.2a; Tansley, Price Evans 1946, pp.48–49.
55. Allen 1986, p.97; Walters 1981, p.93.
56. Cowles HC. 1899. The ecological relations of the vegetation on the sand dunes of Lake Michigan. *Botanical Gazette* **27**, 95–117, 167–202, 281–308, 361–391.
57. Clements FE. 1916. *Plant Succession.* Washington: Carnegie Institute.
58. Clements FE. 1928. *Plant Succession and Plant Indicators.* Washington: Carnegie Institute.
59. Moss, Rankin, Tansley 1910, p.149.
60. Tansley 1947a, p.134.

7 The Managing Director of British Ecology

You not only are the managing director, so to speak, of British ecology, but you are the outstanding (European) figure … and thinker, which is much more important.
Frederic Clements to Arthur Tansley, 12 April 1915[1]

In merely 11 years, between 1902 and 1913, Arthur Tansley moved from relative obscurity to pre-eminence in his profession. Either consciously or unconsciously, and given Tansley's penetrating mind it is difficult to believe the latter, he made four moves that on a chessboard would have entitled him to announce 'check'. First, he founded the *New Phytologist* journal (1902), which gave him control over a large part of what was published by his generation of botanists. Second, he organised and edited the proceedings of the first International Phytogeographic Excursion (IPE I) (1911), so presenting himself to the international community of botanists as the leading British (if not European) exponent of the new discipline that was ecology. Then, third, he helped guide the metamorphosis of the British Vegetation Committee into the first Council of the new British Ecological Society (BES) (1913). In so doing he secured for himself its Presidency, and became within two years the Editor of its journal. The fourth move was slightly different in that it affected not only his professional status – and personal development – but also his family life. It occurred in 1907 when he moved away from Oliver and University College, London (UCL) to a lectureship at Cambridge University. In one sense the move was backwards, for Tansley was moving from a young innovative university to one that was steeped in tradition, but, to compensate, enormous kudos was attached to an appointment at the ancient university. Tansley must have taken pride in the certain knowledge

Shaping Ecology: The Life of Arthur Tansley, First Edition. Peter Ayres.

that his father would have strongly approved of the move. Most significantly for his career, he could in Cambridge be judged 'his own man', separate from Oliver. In Cambridge he could forsake the anatomical studies that had occupied much of his time at UCL and concentrate his attention on ecology. He did so with the energy of a man who had recognised his destiny.

Tansley's progress may in retrospect appear smooth, confident, even imperious, but as each step in that progression is described in this chapter it should be remembered that Tansley was never far from self-doubt. If by the outbreak of World War I in 1914 he had every reason to be pleased with the way his career had developed, his happiness was not complete, for signs of inner dissatisfaction were detectable. By the end of the war, four years later, his problems, both professional and personal, had surfaced, as will be seen in the next chapter.

***

The *New Phytologist* was the first great success of Tansley's career. Just turned 30 years old, and a mere assistant lecturer at UCL, he had recognised that botany and botanists were being held back by a lack of opportunity to publish and discuss their work. What he wanted was a means of communication through which like-minded young men and women could share their knowledge and experiences for mutual benefit but, as he told later, when he had approached other, presumably more senior botanists, he had failed 'to get any satisfactory promise of support'.[2] Undeterred, finding £130 from his own pocket to float the enterprise, and using a cheap printer in a back street off London's Tottenham Court Road (see Figure 3.1), he launched the new journal in 1902. The annual subscription of ten shillings entitled readers to ten, very thin, issues.

Tansley almost called his offspring the *British Botanical Journal* but, following Oliver's suggestion, settled on a more memorable name, *The New Phytologist*.[3] 'New' because a magazine-style *Phytologist* had enjoyed a short but distinguished life between 1842 and 1863. Tansley had wanted 'a medium of easy communication and discussion between British botanists on all matters … including methods of teaching and research'.[4] Publication should be rapid, allowing the swift appearance of new observations otherwise liable to be lost. Short notices of important new books and papers were to be a regular feature, as were short reports of meetings of botanical and biological interest, foreign as well as domestic. The *New Phytologist* was to be *the* forum for botanists.

His instincts were proved right. Although the flow of copy was slow at first – he even had to fill pages in volume one with a letter addressed to himself, as Editor of the *New Phytologist*, sent from himself as assistant lecturer at UCL – the journal soon flourished. Within two years, the *New Phytologist* was paying its way and he had recovered his investment. By 1909 he could with reasonable confidence increase the size of each issue to 36 pages, though he confided to Marie Stopes, 'I only trust one will not be so anxious about one's human children approaching an idea as one is about one's journalistic child'.[5]

The list of authors contributing to early volumes included Oliver and several UCL staff and ex-students, such as Marie Stopes and Ethel Thomas. Edith Chick published there, as did the Blackman brothers, Fritz and Vernon. Although ecology was especially well represented – at least until the *Journal of Ecology* was launched in 1913 – Tansley was adamant that in its pages the *New Phytologist* should include all aspects of botany. To this day, and unusually, the journal remains broad spectrum in its content and appeal. The list of authors slowly expanded, reflecting Tansley's travels and his widening circle of contacts, in particular drawing in those botanists he met through IPE I. Tansley ensured that 'his' journal was where the events of that Excursion were reported, just as he had done earlier with reports of the Erquy expeditions from UCL.

Tansley was a frequent contributor to early volumes, both in his own name, as when he reported his observations on the vegetation of the shoreline in Ceylon, and anonymously as Editor. Very often there was a message, sometimes implicit, sometimes explicit. Thus, while Oliver and Tansley's, 'Methods of surveying vegetation on a large scale', published in the journal in 1904, was presented as an aid to field botanists, its implied message for readers was that these are the methods which you (the field botanist) should follow. It asserted the authors' authority over such matters. Tansley was not afraid to use 'his' journal as a pulpit from which to preach his view of the future of botany. Most famously, in 1917, he stirred controversy when he encouraged the 'Botanical Bolsheviks' to criticise the way botany was taught in British universities, as will be seen in the next chapter.

The promotion of botany through improved communication between botanists was Tansley's aim as he founded the *New Phytologist*. It was a goal he pursued, by various means, throughout his working life.

***

As commonly occurs after meetings of botanists to this day, those of the Vegetation Committee were often followed by field excursions. Purportedly giving the botanists a chance to inspect features peculiar to the local vegetation, the unwritten purpose of such outings is to encourage the forging of friendships and working relationships, while allowing tired and cramped limbs to be stretched in fresh air. Led by Tansley, the Committee members thus visited Crockham Hill in Kent in 1907, where they contrasted the stunted oak, heath, and pines of the higher ground with the stouter oaks of the ancient woodlands on the lower land of the Weald. On another occasion, after a meeting of the Committee linked to the British Association for the Advancement of Science

(BAAS) meeting in Dublin in September 1908, Praeger took his fellow botanists to the bogs and hills of Connemara, and to the Carboniferous limestone region straddling the Clare–Galway border.

The latter meeting underlined for Tansley the benefits of contacts made and the 'amount that could be learned from the examination of vegetation under the guidance of native botanists'.[6] Earlier in the same year he had joined a botanical excursion through Switzerland organised by Professor Schröter of Zürich, an excursion connected to the International Congress of Geography at Geneva. It was in all respects for Tansley a model of what an international excursion should be. It lasted 11 days, and they visited Mount Pilatus near Lucerne, the lakes of the Canton Ticino and the neighbouring part of Italy, the Bernina region, as well as 'the roof of Europe', the area around St Moritz in the Engadine. Among the party were such distinguished European botanists as Ostenfeld from Copenhagen, Rübel from Zurich, and Flahault from Montpellier. Deeply impressed by the value of such meetings, Tansley proposed to the British Vegetation Committee in December 1908 that it should organise an international meeting on an even grander scale. It is testimony to Tansley's powers of persuasion that his ideas were 'favourably received',[7] although, probably, the personal experiences of at least some other members of the Committee may already have convinced them that international links would bring great stimulus to the Committee's work. Indisputably, however, it was Tansley who had seized the moment and who would lead the enterprise.

Turning such a bold idea into reality took, of course, a great deal of work. Letters flowed backwards and forwards between members of the Committee but within 12 months all agreed that they should host an International Phytogeographic Excursion, to be held in August 1911.

Many of the sites of greatest scientific interest were in distant and remote parts of the British islands, so a list of those to be visited was difficult to settle but, bearing in mind constraints imposed by time, ease of rail travel, and availability of accommodation, an itinerary was finally agreed (Figure 7.1). The various constraints meant that, regrettably, only 12 foreign participants could be invited. One declined and, to the organisers' great regret, three more had to drop out at the last moment: they were Warming, Flahault, and C. A. Weber (a great German authority on peat vegetation). Nevertheless, the party that assembled in the Cambridge Botany School on the morning of Tuesday 1 August included such renowned botanists as Drude, Ostenfeld, Rübel, and Schröter from Europe, and Clements and Cowles, both accompanied by their wives, from the USA (Figure 7.2). Male members of the party had been encouraged to bring a 'warm tourist suit and waterproof cloak, with flannelled shirts and strong-nailed thick-soled walking boots'. The women, who each took many

Figure 7.1   The route of the International Phytogeographical Excursion, 1911 (IPE I). (After Sheail 1987.)

photographs during the Excursion, were left to interpret the sartorial advice in whatever way they saw fit but they were warned:

> the British standard of comfort in the smaller hotels and inns is very much below the Continental standard. The beds, though generally clean, are not always comfortable, and the cooking of the food frequently leaves much to be desired.[8]

Figure 7.2   Cowles under 'L' and Clements under 'E' of LUGGAGE at the railway station, Truro, the final stop on the main Excursion. (By permission of the Special Collections Research Center, University of Chicago Library.)

At most sites the vegetation was demonstrated by local botanists, but always present were Tansley and George Claridge Druce, who as author of three county floras had, Tansley acknowledged, an unsurpassed knowledge of British floristic botany. Druce had accumulated considerable personal wealth through his pharmacy in Oxford's High Street and was Secretary of the Botanical Exchange Club, a post he held from 1903 to 1932. Although Tansley probably had a good idea that not all members of the Vegetation Committee liked Druce – Praeger viewed Druce as a 'Good botanist, but puffed up with conceit'[9] – and personal relations between Tansley and Druce deteriorated in later years (Chapter 9), Tansley had had little hesitation in enlisting Druce's help as a guide on the Excursions and in asking him to help sponsor the IPE financially. Druce had had no hesitation in agreeing to both requests.[10]

The programme set aside two days at the beginning to allow participants to get to know each other in the context of a series of social events. On the first day they were guided around Cambridge University's Botanic Garden by the Curator, and then given lunch by the Master and Fellows of Emmanuel College. On the second day, they 'ascended in punts and canoes [along the river] as far as Grantchester', to be entertained to lunch by Mr and Mrs Tansley.[11]

The serious business began next day when, at 6:45 am, the party boarded the train for Norwich. It was the beginning of a strenuous tour lasting four weeks and ending at Portsmouth where the annual BAAS meeting was being held. During those four weeks the party saw the wetlands of the Norfolk Broads and the coast at Blakeney – where their guide, Oliver, showed them his field laboratory and, demonstrating a novel technique, ascended a ladder to record quadrats of vegetation photographically. They were shown by Moss the sharp delineation between the vegetation of the limestone and millstone grit sides of the Derbyshire Dales. In Scotland they saw arctic/alpine vegetation in the Highlands and were entertained by the senior Scottish professors, Isaac Bayley-Balfour (Edinburgh) and Frederick Bower (Glasgow). In Ireland they walked across the limestone terraces of The Burren, guided by Praeger. Even at Portsmouth, their final destination, further field trips were organised for them, with Tansley noting in triumph:

> It was during our visit to Kingley Vale that Drude, whom I had had some difficulty in persuading to come, because, as he said, he wanted to write letters and had seen yew woods before, became tremendously excited and cried 'You did not tell me you were going to show us the finest yew wood in Europe'.[12]

In sentiments unusual in a scientific journal, even in 1911, Tansley was candid:

> A more amiable and harmonious party, a set of people easier to deal with, more anxious to be pleased and to make the best of everything, can hardly be imagined. There was absolutely no friction of any kind whatever, and many old friendships have been strengthened and new ones formed.[13]

His feelings were reciprocated by the Committee's 'guests'. In the *New Phytologist*, Cowles expressed his gratitude and 'admiration for the splendid organisation of the British plant geographers'. Adding:

> The hearty greetings that met us everywhere, the interest expressed in our work on the part even of those who could know little of it, the constant display of whole-souled British hospitality – all these things and many more have made August 1911 a most memorable period in our lives.[14]

The hospitality was, indeed, lavish for Tansley had raised more sponsorship money than was strictly needed, freeing him to plan at each overnight stop extravagant dinner menus, complemented (after his own heart) by extensive wine lists. 'Guests' had only to pay a small fee and bring evening dress, as well as dark suits, for the dinners. After dinner at the Royal Hotel in Truro on the final night of the main Excursion, the party enjoyed communal singing, led by local artistes, and recorded the debt to Druce by signing his dinner card with grateful and affectionate remarks such as 'for kind remembering' or 'with love'.[15] Druce continued to receive letters of appreciation after the guests had returned home. The complimentary reviews that were written for the *New Phytologist* were compiled by Tansley into an expensively bound leather folder which he gave to Druce in gratitude for his support.

The Excursion had been an unqualified success on a personal level. As Clements wrote to Tansley after his return to Minnesota:

> Mrs Clements and I recall with much vividness and great pleasure the days spent with Mrs Tansley and you at Grantchester. Mrs Clements joins me in the very kindest of regards to both of you.
>
> *F. E. Clements to A. G. Tansley, 17 November 1911*[16]

In the same letter, Clements added, 'the amount and kind of ecological work that "you-all" [*sic*] are doing in Britain was astounding, and I am planning on keeping much more in touch with it than in the past'. While seeing certain types of vegetation for the first time, Cowles had been impressed by the 'vast amount of wild country in densely populated England'. He was especially pleased to observe successional series, 'whose progression or retrogression was quite as evident as in American formations'.[17] The latter remark was a subtle criticism of Clements for the pair disagreed about successions, as was revealed when Cowles (1919) reviewed for the *Botanical Gazette* Clements' book, *Plant Succession* (1916).[18] Whereas Clements believed 'succession is inherently and inevitably progressive', Cowles argued it could be retrogressive as well as progressive, though the latter was more common and important. (As an example of a retrogressive succession, Cowles cites an area that is gradually sinking so there is a gradual retrogression from a climax mesophytic forest to a hydrophytic association.)

Cowles had become convinced that international meetings could effect a reduction in 'the misunderstanding of viewpoints, and in the amount of polemic literature'. So highly did Clements and Cowles value their experience in Britain that before the trip ended they pledged to organise a similar excursion, IPE II, in North America as soon as possible.

The second IPE proved even longer then the first. The party gathered in New York City on Sunday 27 July 1913. Starting with the heath-like pine barrens of Long Island and the salt marshes of New Jersey, the Excursion took in close to a hundred sites of botanical interest before it ended at the beginning of October. The ten European guests included Rübel, Schröter, and both Mr and Mrs Tansley, Edith not joining the main party until it reached Chicago on 1 August (Figure 7.3).[19] Chicago was used as a base from which the Excursion could take a series of relatively short journeys to see the 60-metre high dunes at Dune Park, a beech–maple forest at Three Oaks, an example of virgin deciduous forest, and the different stages in the formation of bogs at Mineral Springs – all sites so familiar to Cowles that he could almost count them his own. He was, by all accounts, 'in his glory' (Figure 7.4).[20]

A highlight for the party was when they reached Minnehaha, 2540 metres high in the Colorado Rockies (the Manitou and Pikes Peak Cog Railway has run up the canyon since 1891, but Tansley claims the last 560 metres were ascended on foot). They were in the midst of 'practically untouched forest and

Figure 7.3 Edith and Arthur Tansley, at the University of Chicago, during the International Phytogeographical Excursion, 1913 (IPE II). (By permission of the Special Collections Research Center, University of Chicago Library.)

Figure 7.4 Cowles and Tansley in the field during IPE II. (By permission of the Special Collections Research Center, University of Chicago Library.)

with a glorious view down the canyon over the Great Plains'.[21] This was where Clements had been steadily building since 1900 his summer research base, what he liked to call his 'Alpine Laboratory', on the slopes of Pikes Peak. The party was housed in two wooden bungalows owned by the Clementses and also at the local inn. On their site visits they benefited from Frederic and Edith Clements' encyclopaedic knowledge of the region, an area rich in lakes and bogs, dense spruce forests and open aspen woodlands.[22]

On a personal level, the Tansleys were able to renew their friendships with Frederic and Edith – an inseparable couple whose work together was interrupted only once in 46 years of marriage[23] – staying with them in their own bungalow. The two men shared an admiration for Herbert Spencer, in whose *Principles of Biology* (1864) are ideas that may have helped shape Clements' superorganism concept (Chapter 8),[24] but they were in most respects very different, so Tansley's liking for Clements is at first sight surprising. While Tansley enjoyed good food and wine, and was inseparable from his pipe, Clements was, in Tansley's view, 'decidedly puritan, even ascetic (he neither drank nor smoked, and it gave him real pain to see other people doing so)', but he had in Tansley's eyes redeeming qualities, 'He was essentially reasonable in argument, and he had that best of all senses of humour which enables a man to laugh at himself. He was most kindly and considerate in all personal relationships'.[25] Clements must have found similar qualities in Tansley; in Edith Clements' words, Frederic saw Tansley as, 'a worthy opponent in matching knowledge and intellect'.[26] Although both Edith Tansley and Edith Clements were botanists, their friendship, which began during IPE I, was deep and based on more than one common interest. Like their husbands, they continued to exchange the warmest of personal letters for many years after IPE II.

In some ways, the experienced European botanists were akin to students seeing for the first time plants and communities they had only read about in books. They did not, however, leave their critical faculties at home for they contributed to often lively discussions about the course of succession in local vegetation, sometimes forcefully disagreeing with their hosts.[27] The summer was long and hot. In Nebraska, where Tansley had the honour of meeting the grand old American botanist, Charles Bessey, temperatures reached 108°F as they inspected the high prairies dominated by *Stipa spartea* (porcupine grass). At the appropriately named Mecca, in southern California, 'a thermometer showed 115°F'.[28] The lengthy itinerary included the Great Plains, the Rockies, the redwood forests of California, and, finally, the desert flora of Arizona and the Mexican borderlands. Four National Parks were included along the way. The Excursion was occasionally punctuated by some applied botany:

> On September 5th the party returned to Medford [Oregon] and were entertained to dinner by the Medford Commercial Club and the University Club of Medford – the ladies of the party by the Women's University Club. On the following morning some of the famous pear orchards, beautifully kept and cultivated, were visited under the guidance of their owners. ... The air of happy and abounding prosperity which this rapid and overwhelming success has given to the town of Medford is very striking ... everyone is bright, cheerful and confidant.[29]

and often by good eating, drinking, and relaxation:

> The party received a warm welcome at Salt Lake City, where they were entertained
> to lunch – prefaced by a specially excellent cocktail – at the Commercial Club, the
> remainder of the day being spent in sight-seeing, including a trip to Saltair Beach,
> where several members of the party enjoyed the peculiar experience of bathing in
> the Lake.[30]

In mid-September, the Tansleys left what was probably by now an exhausted
party, the couple heading back to Chicago for a few days rest before returning
home by way of Boston.

As Clements had written after IPE I,

> Without doubt, the greatest personal return from the Excursion was the first-hand
> insight into the point of view of ecologists from different countries, and the
> chance thus afforded of scrutinizing one's own concepts in the light obtained.[31]

In his lengthy account of IPE II, written for the *New Phytologist*, Tansley ended,
somewhat enviously:

> In the vast field of ecology America has secured a commanding position and
> from the energy and spirit with which the subject is being pursued by very
> numerous workers and in its most varied aspects, there can be little doubt that
> her present pre-eminence in this branch of biology … will be maintained.[32]

By contrast, from Clements' side of the Atlantic the prospects for ecology did
not look so rosy. In continuation of the letter quoted at the head of this chapter,
he wrote to Tansley in 1915:

> Certainly if ecology is to go through a long period of confusion, such as taxonomy
> in this country at least finds itself, then some of us will want to find new fields for
> work. It will go through just such an experience unless a few of us do everything
> in our power to prevent it. Your power in this matter is much greater than mine
> because you are in the midst of a thoroughly well organised situation, in which
> you have what seems to the outsider like a controlling or at least a guiding hand.[33]

Further excursions, planned for 1915 by the Swiss phytogeographers who
attended IPE II, had to be postponed when World War I spread across Europe.

As worldwide interest in vegetation studies had grown steadily during the
late 19th and early 20th centuries, it was almost inevitable that internationalisation should occur, with investigators increasingly feeling the need to validate
their own experiences and conclusions, testing the limits of their applicability
to other situations.[34] Tansley's genius was to sense among this community of

scientists a receptiveness to the idea of internationalisation and their need to discuss their differences. He took the lead, organising and facilitating personal interactions that, even interrupted by war, would develop a forward momentum no longer needing his direct involvement. A rift did appear, however, between, on the one hand, Continental plant sociology with its emphasis on the floristic characters of vegetation and the recognition of spatial patterns in which individual species occur, and, on the other, Anglo-American ecology with its emphasis on the dynamics of vegetation and the role of the environment. The rift was greatly regretted by both Tansley and Clements, but their attempts to stop its growth met with only limited success (Chapter 12).

To return briefly to IPE I, its immediate and tangible outcome was Tansley's *Types of British Vegetation* (1911b), a book which the organising committee hoped would have ' the character of a guide to British plant associations ... [but] wider in its scope than a guide for this excursion alone'.[35] Dedicated 'To Professor Eugenius Warming the father of modern plant ecology and to Professor Charles Flahault who through his pupil Robert Smith inspired the botanical survey of this country', the book was, as Tansley wrote in the frontispiece, 'the first attempt at a scientific description of British vegetation'. Tansley insisted that the definite article should be omitted from the book's title in order to avoid any suggestion that *all* existing types were included. Advance copies were available at the start of the Excursion and remaining copies of the limited print run were sold at a break-even price of six shillings. They soon sold out.

The other six members of the organising committee each contributed at least one section, but Tansley wrote the most and, with the help of Moss and W. G. Smith, edited the whole. Tansley's preface is revealing. The book was, he said, 'An endeavour ... to recognise and describe the different types of plant community existing in the natural vegetation of these islands, and to trace their relations, so far as these have been elucidated, to climate and soil, and to one another'. He noted, 'The work of systematically surveying vegetation and recording the results on vegetation maps was begun in Scotland by the late Robert Smith ... and continued by his brother, WG Smith and various other workers'. 'The memoirs and vegetation maps published by these workers ... have formed the nucleus ... of the present book, which is thus a direct outcome of the work of the Central Committee [Vegetation Committee]'.

In the first chapter, 'Introduction, the units of vegetation – their relationships and classification', those units were defined by Tansley. On safe grounds, and over-simplistically from the viewpoint of the IPE members, Tansley told his readers, 'The list of species, arranged taxonomically, is called the *flora* of the region or country, and the study of their distribution is *floristic* plant geography'. The fundamental units of vegetation are wood, moor, heath, marsh, etc., and a unit always develops in a habitat of definite characteristics (Box 7.1).[36]

Box 7.1   How units of vegetation are treated in *Types of British Vegetation*

In their more obvious forms, the units of vegetation have common names in all languages, for example, in English, wood, moor, heath, and marsh. Most of these names refer to something more than vegetation alone, thus 'marsh' implies wetness, 'moor' implies certain sorts of plants but also the presence of a peat soil upon which those plants flourish. In Britain there are eight such categories: woodland, grassland, heathland, moorland, fenland, maritime, sand dunes, and shingle beaches.[37]

From the opposite standpoint, synthetic rather than analytical, certain kinds of plants are always found associated together under definite conditions of life and may be called plant communities (or units of vegetation). The largest of these units, the plant formation, occupies a habitat with constant general characters, i.e. ecological factors which may be climatic or edaphic. Some formations do correspond to a physiognomic group, for example:

*The heath formation** (scrubby low-growing vegetation, usually evergreen)

- Oak–birch association
- Heath association (treeless)
- Pinewood association

While other formations cut across such superficial classification based on physiognomy, or mere topography, for example:

*The vegetation of calcareous soils**

1. Sub-formation of older limestones:
   - Ashwood association
   - Limestone scrub association
   - Limestone grassland association
   - Limestone pavement association
2. Sub-formation of chalk:
   - Chalk beechwood association
   - Ashwood association
   - Chalk scrub association
   - Chalk grassland association
3. Sub-formation of marls and calcareous sandstones:
   - Ash–oakwood association
   - Scrub and grassland association

* See, respectively, chapters 4 and 6 of *Types of British Vegetation* (Tansley 1911b).

'The characters, or as they are often called the *ecological factors*, of the habitat, which influence vegetation, are often classed as *climatic* and *edaphic*.[38] Leaning heavily on Moss's paper, 'The fundamental units of vegetation'(1910), and risking the disapproval of European botanists (Chapter 6), Tansley outlined the British Vegetation Committee's view that the *formation* was the natural vegetation occupying a habitat with constant general characters, while an *association* was the next vegetation unit down in a hierarchy: in practice, it was the community we recognise in the field, usually dominated by one or a few species. Moss had stressed the developmental or successional nature of vegetation and to emphasise the importance of the dynamic nature of vegetation – a perspective that eventually led Tansley to conceive the ecosystem as 'a position of relative equilibrium'[39] – Tansley briefly outlined the work of one of the IPE's guests, Clements, on primary succession of associations (called formations by Clements).

After IPE I, Cowles, for one, conceded that the British analysis of vegetation was the most workable yet proposed.[40] There was no coherent opposition. As sagely pointed out by Tansley many years later, 'It is not really significant for the progress of our knowledge of vegetation how the term 'formation' is used (or whether it is used at all), provided we know the sense in which a particular author understands it'.[41]

Part I of *Types of British Vegetation* comprised two chapters describing the climate and soils of Britain, while the 14 chapters of Part II focused on the sites to be visited. Thus, W. G. Smith wrote about Scottish heaths and Moss wrote about the Pennines. Frank Oliver wrote about the shingle beach communities of Blakeney, while Marietta Pallis described her studies of the Norfolk Broads and fen formation, work that she had had no opportunity to publish elsewhere.

The print run of the book was rapidly sold out and Tansley soon contemplated a revision. However, more than a quarter of century would pass and ecology would come of age before, drawing upon his lifetime's experience, he produced an effective replacement, *The British Islands and their Vegetation* (1939a).

***

At this point a short digression is needed to allow us to catch up with some relevant events in Tansley's personal life. On 1 March 1904 King Edward VII opened a new Botany School in Cambridge. As Tansley remarked to Marie Stopes two months later, as she was about to return from studying in Munich, 'I fear you will find UCL terribly dirty and cramped in spite of its being "home". The new Cambridge Institute [School] is far the best place now so far as facilities and space are concerned. It tempts one to move to Cambridge'.[42] He succumbed to that temptation in the autumn of 1906, moving from UCL to Cambridge when a lectureship became available after the incumbent professor,

Harry Marshall Ward, died. The chair thus vacated was filled by one of the Botany School's lecturers, Albert Charles Seward. By the time of their move out of London, Arthur and Edith Tansley had a growing family. Their first daughter, Katharine, had been born in 1904, and Margaret followed the next year. A third daughter, Helen, was born in 1909.[43] The Tansleys purchased Grove Cottage in Grantchester, a village less than three miles south of Cambridge city centre. The large house could be easily modified to suit their needs and they loved its extensive gardens. Clearly, the meagre academic salary that Arthur had enjoyed at UCL would not have financed such a purchase, so it seems likely that he drew in some way on an inheritance from his father, who had died in 1902. The Tansleys remained at Grove Cottage for the rest of their long lives, although for ten years from 1927, when their children were grown up, and following his appointment to the Sherardian Chair in Oxford, Arthur lived away during term times.

The move to Cambridge enabled Tansley to concentrate on ecology. He was soon elected to the governing body, the Syndicate, of the university's Botanic Garden (and was its Secretary from 1910 to 1915). The Garden's Curator had in 1900 made 'new beds ... to illustrate the vegetation characteristic of different kinds of soil, with special reference to plants met with in the Chalk districts and at the Seaside' and by 1908 he was referring to these as the 'Ecological Beds'. Whether or not the change of name betrays Tansley's influence, or possibly that of Moss, the beds soon became the site of experiments laid out by Tansley's students (Chapter 8).[44]

Not everything in Cambridge was an improvement on UCL. In the congested space of the Botany School Tansley was allocated only 'a segment of the elementary laboratory cut off by matchboarding and chickenwire' in which to write his papers and to edit the *New Phytologist*.[45] However, just as intellectually he had begun to distance himself from Oliver when he founded that journal, so in moving away from London he completed his separation from Oliver, who might otherwise have been thought of always as Tansley's father figure and guide. There is no suggestion that Tansley and Oliver ever enjoyed anything other than the warmest personal relations and, in some ways Tansley was more 'grown up' than Oliver for, whereas Tansley frequently displayed a patient steely resolve in attaining his goals, Oliver retained a boyish enthusiasm untrammelled by ambition. In his studies at Blakeney, for example, Oliver was more interested in the work itself than in its ultimate outcome. A considerable number of papers resulted but much of the research remained unpublished.[46] As Oliver admitted himself, his botanical career was unashamedly erratic, starting with plant physiology before moving to palaeobotany and, finally, ecology. The last twist occurred at the relatively advanced age of 65 when in 1929 he retired from UCL and, after years of studying coastal communities, turned his attention to deserts. Wholly in character, he surprised friends and colleagues by taking on a fresh challenge as he accepted a chair at the Egyptian University in Cairo, which he held for the next five years.

***

Change was in the air in the spring of 1912:

> it was felt that the original stream of ideas which informed the first labours of the [Vegetation] Committee had been – not indeed exhausted – but rather widened out into many and diverse channels, and had been joined by other ideas and lines of study contributed by the accession of new workers and by the general progress of the science. ... the Committee was no longer well-fitted to represent and to guide effectively, ... the time had come to replace the Committee by an organisation of more extensive scope such as might be furnished by a British Ecological Society.[47]

These are the words with which, on the very first page of the first volume of the *Journal of Ecology*, published in March 1913, Tansley explained the origins of the BES. The opinions of potential members on the creation of a specialist society had been canvassed, by way of a circular, and by the time the Vegetation Committee met in December 1912 it was clear that the idea was strongly supported. During two days there was lengthy and often tedious debate about the rules of the new Society but, finally, the Committee unanimously resolved 'to create a Society to take its place and carry about its work'. A council of 12 members, including a President and two Vice-Presidents, would govern the Society; those 12 in the first instance being simply the current members of the Vegetation Committee.[48] Tansley would be the first President.

Fundamental to plans for the new Society was the publication of its own journal, its importance being demonstrated by the appearance of the first issue of the *Journal of Ecology* in March 1913, one month *before*, on 12 April, the new Society was formally constituted and the old Committee was laid down. Clearly, the winter of 1912–1913 must have been one of intense activity and commitment for all those involved. Since Tansley was to hold the onerous position of President for two years, the journal would be edited by Frank Cavers – who had already proved his worth as Tansley's assistant editor on the *New Phytologist*. Cavers was, in Tansley's view, 'an extremely hard, untiring worker', showing as Editor 'conspicuous powers of masterly condensation and clear exposition'.[49] Tellingly, it was Tansley, not Cavers, who wrote the *Journal of Ecology*'s first editorial, 'The aims of the new journal', from which the extract above is taken. Tansley was pulling the strings, and Cavers would step down when Tansley relinquished the presidency.

In publishing, there is an unavoidable period between the receipt of a manuscript and its final appearance in print – even if no corrections are required – so, for a March publication, Tansley must have been soliciting copy *before* the Committee decided formally to establish a new society and a new journal. One possibility is that, to fill the *Journal of Ecology*'s pages, he switched into it some material originally destined for his *New Phytologist*. After 1913 he certainly used the *New Phytologist* much less as the newsletter for ecologists and the *Journal of Ecology* much more.

The two aims of the *Journal of Ecology* were, according to Tansley, 'to foster and promote in all ways the study of ecology in these islands', and, through critical articles, reviews, and commentaries, to present 'a record of and commentary on the progress of ecology throughout the world'.[50] The second aim had not been included in the circular sent to potential members. Tansley was stretching the outline approval given by members of the Committee, leading the journal in a direction he wanted it to go. Possibly to forestall any criticism, he explained in the editorial how the new journal would be sectionalised, with work relating to the British Isles appearing first in each quarterly issue, but how inclusion of foreign work would, apart from enriching the subject, enrich the coffers of the BES through greatly increased sales.[51]

As reported by Frank Cavers, who was Secretary of the Society as well as the Editor of its journal, at the annual general meeting (AGM) held at UCL on 23 May 1914, the journal lost money in its first year. This was not unexpected because many complimentary copies had been printed and circulated for publicity purposes. Encouragingly, more papers were being submitted than could be published and it was hoped that, as membership and subscription income increased, the size of the journal could be doubled to accommodate some of these. 'The Secretary appealed to all Members to assist in every possible way in bringing the Society and *Journal* to the notice of others'.[52]

At the same AGM, Tansley began his Presidential Address appropriately and momentously, 'this is the first presidential address to the first ecological society – so far as my knowledge goes – the world has seen'.[53] After congratulating the Society 'on a successful opening year', in which 'Our numbers are markedly higher than it was estimated at the outset', he moved almost immediately, and revealingly, to talk about the journal, 'the chief, though it is by no means the only, advantage the Society has to offer its members'. The other advantages, he quickly added, included the privileges of attending the meetings and excursions. 'I spent the whole of last summer [1913] in the United States, and wherever I met plant-ecologists [*sic*], who are far more numerous among professional botanists in that country than is the case in the British Isles, I was greeted with the heartiest and most welcome congratulations, not only on the foundation of this Society, but especially on the excellent start made by our Journal'.[54]

The President then used his platform to review the state of ecology, particularly at home in Britain, and to lay out his very personal vision of the way forward. His themes were either familiar to his listeners, or, as when he argued for the value of 'well done' science that was outside 'the main current', would become familiar as he returned to them through his long career.

There was, he said, a need for a school of physiological plant ecology because 'The most fundamental problems of ecology are of course physiological problems'. Outstanding work being done at Rothamsted Experimental Station in Britain and by the US Department of Agriculture was demonstrating the benefits of an ecophysiological approach, for 'agricultural problems are really ecological problems, though of a specialised character'.[55] Ecologists needed to work together with physiologists. While 'we badly want a series of autecological monographs based on the field study of the more important members of the British flora, particularly the dominants in associations', synecological studies of the relations 'between the members of our well-known closed, relatively stable plant-associations' were equally important. Good quantitative data were needed, not endless lists of the ground flora within single woods but data that either compared several woods, or data that related variation in the ground flora of just one wood to variations in, for example, soil type.

Reminding his audience of the importance of competition, he described classic studies made 50 years earlier by Nägeli on the effects of soil type on the distribution of two alpine species, *Achillea moschata*, favoured by siliceous, non-calcareous soil and *A. atrata*, favoured by calcareous soil; studies which foreshadowed his own (unmentioned) studies of *Galium* spp. Pointing to the work of Oliver and colleagues at Blakeney Point, he gave another reminder: not all associations are stable.

Tansley was steering the young science, as was his habit. Given his eminence, its practitioners would be wise to take heed. The Society and its President's long address demonstrated that ecology now existed as a distinct science within biology. It might be riddled with self-doubt, but the President had identified clear pathways forward. As he finally remarked, 'Primary survey will always remain an excellent, perhaps the best possible, training in the general study of vegetation', but, his audience could infer, the future lay with ecology. 'The meeting was concluded by Prof. F.W. Oliver [Vice-President] proposing a vote of thanks to the President for his valuable and stimulating address'.[56]

Metamorphosis of the British Vegetation Committee into the British Ecological Society, with Tansley at its helm, was complete. The original Committee had been changing gradually – Smith and Hardy had left, Praeger had turned his attention back to floristics, and the altered career paths of Moss, Rankin, and Woodhead allowed them much less time for its affairs. Significantly,

Oliver had been added. But the final step was a triumph for Tansley. Through the IPEs he had established himself in the eyes of overseas ecologists as the leader of British ecology. A few months short of his 42nd birthday, his eminence had now been formally recognised by his fellow countrymen as they made him President of the world's first society for professional ecologists. Through his editorship of both the *New Phytologist* and the *Journal of Ecology*, he effectively controlled what was published by British ecologists. He could decide also which papers to publish from among those submitted by foreign ecologists.

Although the subjects of research were not dictated by funding bodies as happens today – making young researchers of that time freer – they would have been wise, if they wanted to get their work published, to choose a subject, and also a methodology, of which Tansley approved. Through his Presidential Address and his frequent editorials in both of 'his' journals, Tansley had made it crystal clear what their priorities should be.

By the summer of 1914 Tansley was well on his way towards shaping ecology to fit his vision. Yet, as proud of his achievements as he must have been, success had not brought complete happiness. There were nagging doubts in his mind, both professional and personal. In the summer of 1914 he had managed to leave those cares behind, however, to do what he enjoyed most. He was leading a party of Cambridge botanists in Provence when news filtered through of the assassination on 28 June of Archduke Franz Ferdinand, heir to the Austro-Hungarian throne. Six weeks later, on 4 August, Britain declared war on Germany. Tansley's problems were about to intensify.

Notes

1. H29 Tansley Archives, University of Cambridge Library.
2. Godwin 1985a, p.1.
3. Currently, and throughout this book, referred to as the *New Phytologist*, '*The*' having been dropped to facilitate library indexing.
4. Tansley 1902.
5. British Library, ref 58468 ff 172.
6. Tansley 1947a, p.134.
7. Ibid.
8. Sheail 1987, p.32.
9. Allen 1986, p. 200.
10. Anker 2001, p.17.
11. Tansley 1911a, p.276.
12. Tansley 1947a, p.135.
13. Tansley 1911a, p.290.
14. Cowles 1912, p.26.

15. Anker 2001, p.18.
16. H26 Tansley Archives, University of Cambridge Library.
17. Cowles 1912, p.26.
18. Cowles must have been hurt when he read Clements' book, *Plant Succession* (1916), for in it Cowles was lumped together with 'previous workers', probably because his pioneering work on succession in the Indiana dunes was then more than 15 years old (Cassidy 2007, p.55).
19. Tansley 1913a, p.323.
20. Cassidy 2007, p.52.
21. Tansley 1914a, pp.30–1.
22. Clements 1960, p.16.
23. Langenheim 1996, p.5.
24. Sheail 1987, pp.19–20.
25. Tansley 1947b, p.196.
26. H28 Tansley Archives, University of Cambridge Library.
27. Tansley 1913a, p.334; Tansley 1914a, p.30–31.
28. Tansley 1914a, p.328.
29. Ibid., pp.270–1.
30. Ibid., p.84.
31. Clements 1912, p.179.
32. Tansley 1914a, p.333.
33. H29 Tansley Archives, University of Cambridge Library.
34. Fishedick, Shinn 1993, p.108.
35. Sheail 1987, p.34.
36. Tansley 1911b, p.11.
37. Tansley 1939a, p.71.
38. Tansley 1911b, p.4.
39. Tansley 1939a, p.vi.
40. Cassidy 2007, p.280.
41. Tansley 1939a, p.viii.
42. British Library, ref 58468 ff 162.
43. Thinking his daughters were 'getting too much God' at their junior school, Tansley gave them lessons in simple science and astronomy. He was successful. After going on to The Perse secondary school in Cambridge and then to UCL, his daughters, like their Chick aunts, built successful professional careers, Katherine in physiology, Margaret in architecture, and Helen in economics (Martin Tomlinson personal communication).
44. Walters 1981, p.92.
45. Godwin 1985b, p.50.
46. Salisbury 1952, p.235.
47. Tansley 1913b, p.1.
48. Members of the first Council of the British Ecological Society were W. B. Crump, O. V. Darbyshire, T. G. Hill, C. E. Moss, F. W. Oliver (Vice-President), R. L. Praeger, W. G. Smith (Vice-President), A. G. Tansley (President), H. B. Watt (Treasurer),

F. E. Weiss, T. W. Woodhead, and R. H. Yapp. Tansley and Yapp tabled the resolution that the Society should be governed by a council of 12 members, two of whom would retire annually and not be available for immediate re-election.

49. Sheail 1987, p.41.
50. Tansley 1913b, p.1.
51. Membership had grown by 1917 to little more than 100. It remained small until after World War I. The Society survived largely on library sales of the journal (Salisbury1964, p.17).
52. Tansley 1914b, p.194.
53. Following discussions organised by H. C. Cowles in 1914, the Ecological Society of America was officially inaugurated on 28 December 1915.
54. Tansley 1914b, p.195.
55. Ibid., p.196.
56. Ibid, p.202.

8 Disillusion and Disaffection

Arthur Tansley combined high intellect with an open, fertile mind. Philosophy and, especially, psychology vied with botany for his attention. While he managed to contain the former largely as a hobby, or as an occasional adjunct to his science, psychology sometimes predominated, its precepts even enriching his botanical thinking. Botany may have been his priority through most of his life but there was a period, from the early 1910s through to 1927, when disillusion with botany and botanists caused such disaffection that he was increasingly tempted to give up the subject in favour of a career in psychology. That period of unhappiness is the subject of this chapter.

The robust health of ecology, and of ecological societies around the world, is nowadays commonplace and unremarkable. The same was not true a century ago. Tansley may have been 'an acknowledged leader of ecological thought and progress' but, as Godwin remembered, 'his kingdom was still extremely small … not a few thought it a visionary one soon to be dissolved by the impact of severer scientific reality'.[1] An especially low point for British ecology was reached in 1921. Growth in the membership of the British Ecological Society (BES) had stalled, and its Summer Excursion – a residential meeting planned to be held in Taunton – had to be cancelled because of lack of interest.[2] Day excursions organised for members in the London, Manchester, and Cambridge regions were also cancelled, ostensibly because strikes by coalminers were making rail journeys unreliable. Only seven research papers were published in the BES's journal in that year, the remaining pages being filled with notices and book reviews. The future of ecology was far from rosy, and inextricably linked to its fortunes were Tansley's own chances of advancement.

Shaping Ecology: The Life of Arthur Tansley, First Edition. Peter Ayres.
© 2012 by John Wiley & Sons, Ltd. Published 2012 by John Wiley & Sons, Ltd.

The BES's struggles to shake off the after-effects of war would have depressed Tansley's mood, but his period of disillusion and disaffection had begun much earlier. As he celebrated his birthday on 15 August 1910, entering his 40th year and mid-life, he was quite naturally becoming increasingly concerned about his own position. He was still plain 'Mr Tansley', a commonplace university lecturer whose talents and achievements were, it must have seemed in his darker moments, unrecognised outside his immediate circle of colleagues and contacts. However, after his good friend Vernon Blackman wrote to him on 10 November, it looked as though his peers had, after all, been following his progress and appreciating his successes. They might be about to elevate his status with the award of the highest scientific accolade, a Fellowship of the Royal Society of London. Blackman's letter began, 'My dear Arthur, I heard from Oliver a few days ago there was some chance of you being put up … [for a Fellowship of the Royal Society] … you ought to stand'.[3]

Any raising of Tansley's spirits must instantly have been dampened as he read on, for Blackman continued, 'I cannot say that I approve of the new strategy, as Oliver calls it, which is to enlarge the waiting list so that occasionally two botanists may get in at once'. Oliver, who was already a Fellow of the Royal Society and therefore on the inside, had suggested that Vernon Blackman himself should also 'go forward', along with William Lang and David Gwynne-Vaughan, both of whom were well-known to Tansley within the small botanical world of 1910. Blackman passed on Oliver's warning, 'with 100 candidates each year, it is quite common to wait, 2, 3, or even 4 years.' In Tansley's case the wait proved to be closer to five years, and must have been all the harder to bear as he saw Lang elected a Fellow in 1911, and Vernon Blackman elected in 1913. Lang was three years younger than Tansley, Blackman a few months younger. Both already held chairs, Lang in Manchester, and Blackman in Leeds. Perhaps more significantly, both specialised in areas of botany – Lang in palaeobotany and Blackman in mycology – long established and familiar to the Royal Society's members.

It would have been all too easy for Tansley to infer that his involvement with ecology was holding him back, and he would have been correct. Correct because two of the most senior botanists in the Royal Society, Isaac Bayley Balfour (elected 1884) and Frederick Orpen Bower (elected 1891), were becoming increasingly antagonistic towards the upstart that was ecology and, in Balfour's case, to Tansley himself. Balfour, together with Percy Groom, had in 1909 been one of the principal translators into English of Warming's great text *Oecology of Plants* (Chapter 2). A systematist and to some extent a plant geographer himself, Balfour seems to have resented the way that Tansley had assumed leadership over a broad swathe of botany in which he, Balfour, might have expected to be the authority. He disapproved further because Tansley, in support of Clements, had argued for a community-based, synecological, approach to ecology. Bower's support was for the traditional areas of botany; he backed Lang and, in addition, Gwynne-Vaughan, both of whom shared his own interests in palaeobotany and the origins of the land flora. He had worked closely with both men for several years, developing lasting friendships.

When Gwynne-Vaughan learned that Bower was drawing up a certificate for his election, he modestly protested that there was a whole crowd of botanists who should get the chance before him, for example 'Tansley – without hesitation'.[4] Unlike Lang, Gwynne-Vaughan never became a Fellow for tragically he died in 1915, only months after taking a chair in the college that would eventually become the University of Reading.

Tansley was finally elected, however, in 1915. He would always thereafter proudly add the letters 'FRS' to his signature. Now, plain 'Mr Tansley' had an adornment.[5]

His long wait for a Fellowship was just one of several concerns on Tansley's mind when in May 1912 he composed a letter of application for the Professorship of Botany in the University of Sydney.[6] The letter was never sent but his frustrations at the time are apparent in his handwritten draft. As an assistant professor at University College, London (UCL) he had taught, 'Anatomy, reproduction, floral biology, the algae, and the morphology of flowering plants', a range which it may be inferred was quite fairly expected of a junior. However, 'Since coming to Cambridge in 1907', he wrote in the letter, 'my advanced teaching has extended over a wider range than is usual for a single member of staff, including special morphology of mosses and ferns, physiological and morphological anatomy, plant ecology, and certain aspects of reproduction'. Clearly, he felt he was overloaded and the move from UCL to Cambridge had not allowed him to get away from teaching older subjects, like comparative morphology, as he had hoped. A marginal note to his draft letter added, 'I have also suggested directed or given advice upon numerous investigations by post graduate students of this department'.

In preparing his application for the Sydney chair, Tansley had secured the written support of the Blackman brothers, Vernon and Frederick ('F.F.'), and of Oliver. His current professor, Albert Seward, had also promised support, as had Francis Darwin (who was a reader in the same department). Luckily the letters from F. F. Blackman and Oliver survive.[7] As might be expected, both emphasised the loss to British botany that Tansley's departure would represent. For example, Blackman judged Tansley, 'a man of unusual brain-power, of wide interests and many fine qualities of mind'. He noted that, 'recently Mr Tansley has taken up the new branch of botany known as ecology and indeed he is largely responsible for the rapid development of it in this country'. Near the end of his reference, Blackman included a revealing sentence, 'It is only owing to financial independence in the past that he has never applied for a professorship or other major post in this country but has continued to hold a minor post which would allow him freedom for his many activities'.

It is not known why Tansley failed to apply formally for the Sydney post; perhaps he was persuaded to stay by the compliments paid to him by those he respected most. It was not just his referees who lavished praise on him. When he had turned to his old friend William Smith for advice, Smith's response was unequivocal,

> whether your going would damage ecology – my answer is undoubtedly *yes*. What I fear most is that affairs would fall into the hands of dogmatic unteachable people who would take things with a high hand and make heretics of any who did not agree with them. In a developing science this is the greatest error … You have always been open to reason when matters came under discussion, and I don't know anyone who has so much the confidence of all concerned … There has always been too that enthusiasm not too much cooled by caution which has led us into doing what luke-warmness and over-caution would not have done.[8]

***

If one aspect of ecology gave Tansley more satisfaction than another then, in the early 1910s, it was the progress made in the study of woodlands (Chapter 6). Moss, Rankin, and he had set the standards when they wrote 'The woodlands of England' (1910). Among the younger generation of botanists, no one was a keener practitioner of Tansley's methods than Robert Stephen Adamson, who had delivered to the Vegetation Committee in 1910 a paper in which he defined 'the physical coefficients of woodland plant societies'.[9] A 25-year-old Edinburgh graduate, Adamson was working in the Botany School, Cambridge, studying a local wood at Gamlingay. Having made those quantitative observations, on site, soon to be so strongly advocated by Tansley in his Presidential Address to the BES (Chapter 7), Adamson was able by 1912 to publish, 'An ecological study of a Cambridgeshire woodland' in the *Journal of the Linnean Society*. After measuring the light intensity and soil water content at the growing sites of individual plants in Gamlingay Wood, he had plotted their specific names on a graph that had light and water as its two axes. When all the points had been plotted, plants of each species were seen to be clustered in their own distinct regions of the graph, each defined by the intersection of a narrow range of light intensities and water contents. It was the clearest demonstration to date of a principle of great ecological significance: each species has its own peculiar requirements for light and water. It was therefore Tansley's great loss when the talented Adamson, having acquired a Cambridge MA, moved to a lectureship at the University of

Manchester. Although they continued to collaborate, most notably in studies of the chalklands of southern England (p. 117), the 160 miles between Cambridge and Manchester prevented that easy day-to-day exchange of ideas and impressions that underpin the closest collaborations.

While Tansley could take some comfort from the thought that Adamson's departure signalled the appointment of a fellow ecologist to a university lectureship, where Adamson could teach the principles laid down by Tansley, the departure of Charles Moss in 1916 was a loss with few compensations. To some extent Moss was already lost to ecology since for several years past he had been reverting to floristic botany, publishing a British flora in 1913.[10] He was nevertheless one of the co-founders of the British Vegetation Committee, someone with whom Tansley had published jointly and someone on hand in Cambridge who could be relied upon to give Tansley a blunt critique of his latest ideas. Although Moss was a hard-working, reliable member of staff, it became necessary for him to seek another job when his income was halved as a result of the wartime shortage of students. His departure was precipitated by the failure of his marriage and subsequent divorce (the latter was considered rather scandalous at the time). So, when a chair at the South African School of Mines and Technology (soon to be the University of Witwatersrand) became available, Moss applied and was appointed.[11]

These losses were as nothing compared with that of Captain Albert Stanley Marsh to a sniper's bullet. On 6 January 1916, Marsh was shot through the heart as he was passing a gap in the trench parapet in the lines of the Somerset Light Infantry near Armentières in northern France. Marsh was only 23 years old at the time of his death and, in strictly botanical terms, had achieved little. A member of Tansley's college, Trinity, Marsh had graduated in 1913 with a first class degree, 'though … he did not get it too easily', his interests being 'too scattered'.[12] He had joined others in the summer of 1913 in studies of the shingle, salt marshes and sand dunes at Holme, in north Norfolk, subsequently completing and publishing his own detailed comparison between the topography and plant communities found there and at nearby Blakeney Point – the first example of such a detailed comparison.[13] He had also had a short note of the anatomy of Cycads published in the *New Phytologist*.[14]

More pertinent to Tansley's career, Marsh had in 1913 taken over from another student, Eleanor Margaret (Margot) Hume, some experiments at the Botanic Garden. Originally suggested by Tansley,[15] these explored the effects of various soil types on competition between the 'lime-avoiding' (calcifuge) *Galium saxatile* (heath bedstraw) and the 'lime-loving' (calcicole) *G. sterneri* (Sterner's bedstraw=*G. sylvestre*). After Marsh's death Tansley took the work forward himself, eventually publishing his findings in the *Journal of Ecology* (1917).[16] He concluded that it was through competition between shoots (for light), rather than between roots (for water or nutrients), that each species suppressed the other when they grew together on soil closest in character to the dominant species' own natural type. Although Tansley sometimes employed experimental treatments in his field studies, and was a strong advocate of experimental ecology, this was the only time in his career that he was involved with experiments at the much smaller scale represented by garden plots.

In writing Marsh's obituary in *New Phytologist*, Tansley displayed a tenderness and warmth he rarely allowed to be seen in public. He wrote how, in the scholarship examination for entry to Trinity, Marsh's 'work in botany was really wonderful for a boy of his age [17 years]';[17] '... when he showed me the letter [granting him an army commission] – he was like a girl with the invitation to her first ball'. He was 'a man who put all of himself, as a man should, into the job he had taken up'; 'And his servant wrote, "He was not only respected but loved." He had quite the same hold on those who knew him well at Cambridge, and, quite apart from his scientific promise, his loss is very bitter to those who loved him'.[18]

No one who lived through World War I was untouched by the personal tragedy brought by each of the millions of deaths.[19] Arthur Tansley never had a son to lose, so was spared the greatest sadness suffered by so many who did. Without suggesting that Marsh was anything approaching a son to him, he was clearly deeply affected by this young man's loss. His death added to Tansley's growing sense of bitterness and frustration.

The Great War came inescapably to Cambridge when huge, tented, army training camps appeared on the colleges' sports fields. With its headquarters at Trinity College, and its beds in the college's grounds, the First Eastern General Hospital received from France hundreds of wounded, maimed, and gassed soldiers; it later expanded onto the fields of King's and Clare Colleges and by 1915 had 1500 beds. For the inhabitants of the city, war brought a host of minor inconveniences. Domestic staff left their employers, either to join the armed forces or to join the industries supporting the war effort. The annual rhythms of university life were disrupted as student recruitment almost ceased and, as seen with Moss, staff incomes suffered. The survival of the fledgling BES was threatened, and aspirations of the embryonic conservation movement were almost extinguished as the world slid into the abyss of war. Like many, Tansley thought the war would soon be won, but by the spring of 1915, as he told Frederic Clements (a staunch supporter of the Allied cause), his optimism had evaporated.[20] Excluded from the fighting by his age and his deformed hand, he sought to serve his country by taking 'a more or less routine clerking post in the Ministry of Munitions'.[21] The exact nature of his service is unknown but it involved him working eight hours a day, six days a week at the Armament Building.[22] His daughter, Margaret, later told her children that her father had been entitled to wear a uniform – though, on one occasion, this had unfortunate consequences. Walking into Trinity College in uniform one day, he met Bertrand Russell, his friend from student days. Russell was an outspoken pacifist and, upon seeing Tansley, ostentatiously turned on his heel and without speaking left the room.[23] A quarter of a century would pass before the two men were reconciled.

***

When, in May 1918, F. O. Bower accused Tansley and others of advocating 'immediate *Botanical Bolshevism*', he was using language that was not just strong but, at the time, highly emotive. Throughout the previous year, Russia had been racked by revolutionary turmoil, the Czar had been deposed and finally, in October, Lenin's Bolshevik faction had seized power. Britain was threatened in two ways, first by the imminent loss of a powerful ally as Russia began to negotiate what became the Brest–Litovsk peace treaty with Germany, and second by the spectre of an infection of ideas – the overthrow of an ancient monarchy and established order by organised working classes. Bower drove his metaphor further, 'In order to secure their own Utopia they [Tansley and others] propose to "subordinate" something which they admit is good in itself. That is the spirit that has ruined Russia, and endangered the future of civilisation.'[24]

This was not language normally found in *New Phytologist*, so what was it all about? Bower was leading a counter attack made by a group, conveniently called the 'Morphologists', against proposals published in December of the previous year in *New Phytologist* under the heading, 'The reconstruction of elementary botanical teaching' (1917). The article was 'signed' by F. F. Blackman, F. W. Oliver, Vernon Blackman, Frederick Keeble, and A. G. Tansley, all experienced university teachers. The Morphologists were, however, all too ready to believe that because the paper was in Tansley's journal its motivation came from him. The paper became known as the 'Tansley Manifesto'.

The Manifesto argued:

> Botany in this country is still largely dominated by the morphological tradition, founded on an attempt to trace phylogenetic relationships of plants, which began as the result of the general acceptance of the doctrine of descent. Elementary teaching (as well as a very large part of advanced teaching) is mainly occupied with the endless facts of structure and their interpretation from the phylogenetic standpoint. Side by side with this there generally goes a discussion which is often limited by a crude Darwinian teleology. Plant physiology is relegated in most cases to a subordinate place and is taught as a separate subject. The newer studies of ecology and genetics play a very small part in the curriculum.[25]

Pouring petrol on their fire, they claimed the current morphological approach did not 'attract the best types of mind amongst possible students'. Nothing could have been more calculated to anger the Morphologists and provoke civil war.

With an eye, no doubt, to increasing readership of the *New Phytologist*, Tansley published not only the original Manifesto but also called for responses to be sent to him for publication in the same journal (as has been seen, among those who responded was Bower). In the eyes of the Morphologists this action merely underlined Tansley's leadership of the 'Manifesto' group. However, when, some 70 years later, the whole affair was analysed objectively by Glasgow University historian, Don Boney, the evidence of published and, often more importantly, unpublished letters suggested that it was the Blackmans, rather than Tansley, who were the instigators of the Manifesto. Tansley had merely redrafted their original outline and kept the controversy going via the pages of his journal.

Tansley, like his brother-in-law, F. F. Blackman, had risked provoking the displeasure of his head of School, Professor Albert Seward – as indeed he did, for Seward judged the Manifesto 'monstrous'[26] – but it was a price worth paying if physiology and ecology were to win their proper place in the botanical curriculum. For Tansley it may have seemed that there was much to gain and little to lose, for he was of independent means, frustrated with the state of botany and, increasingly, attracted by psychology. What he had not foreseen was that the Oxford Chair of Botany would very soon become available. It was to prove a very bad time for him to antagonise the more traditionalist members of his profession.

In the responses of the Morphologists there was as much pedantry as there was substance. Bower may have been their spokesperson but, behind the scenes, opposition was led by Isaac Bayley Balfour: the opening sentence of a personal letter to Bower, dated 17 April 1918, gives a flavour of his mood, 'My dear Bower, You may well ask who makes them lords over us?'[27] Boney concluded that 'the dedicated revolutionaries of the 1870s and early 1880s [such as Bower, who as a young man had studied with Julius von Sachs in Germany], the ardent protagonists of what was then the "New Botany", became in later life … entrenched and inflexible … confirmed conservatives.'[28] Some younger palaeobotanists, such as Lang and Marie Stopes, fought on the side of the Morphologists but Boney found that not only was the overall balance of letters received by Tansley broadly in favour of the Manifesto, more importantly, its supporters came 'from individuals newly appointed to chairs, or who were soon to be professors, many of whom in the next 20–30 years would make significant impacts on those branches of botanical science, physiology and ecology, then under discussion'. The Manifesto had its desired effect.

When the Chair of Botany in Oxford became available later in 1918, Tansley found he had ruffled too many feathers. His application was unsuccessful. As he remarked to Frederic Clements, in a letter dated 18 December 1918, 'I've been getting some experience in the "Gentle art of making enemies" lately … Reactionary forces are pretty strong here, and it will be a hard struggle to get anything progressive done. But I am going to have a good try. In regard to the "reconstruction" discussion [Manifesto] the enemy has had his innings and the end of it will be mainly on my side. … Fortunately my livelihood does not depend upon the favour of the exalted reactionaries'.[29] Although Seward, one of the Oxford electors, was prepared to overlook Tansley's part in the Manifesto affair and back his candidature for the chair, the other electors, I. B. Balfour and D. H. Scott,[30] were not. Indeed, Balfour had injudiciously declared his hand before he realised that he was to be an elector. Writing to Bower on 12 September 1919, he had suggested scathingly that Tansley's measure would be apparent once he was away from Cambridge and its prestige. In other words, Tansley would be found wanting. In the event, Frederick Keeble, Director of the Royal Horticultural Society's Garden at Wisley, was appointed to the chair – which as the next chapter will show was a disaster for Oxford – Balfour and the other electors overlooking the fact that Keeble was one of the authors of the Manifesto.[31]

Tansley's pride had suffered yet another blow. While Bower was fair minded enough to realise this, commenting to Balfour, 'I cannot help feeling sympathy with Tansley in what for him will be a bitter disappointment. I hope it will not sour him or cause any unhappy split', Balfour showed no sympathy. Indeed, he was exultant. Replying to Bower, he wrote, 'I laughed in my sleeves when at the election of the Oxford Professorship Seward, solemnly and at great length, gave as one of Tansley's chief claims to the appointment "his signal services as a reformer on botanical teaching"'.[32]

Further light is cast on Tansley's standing in the botanical community by another spat that occurred a few years later. Although Tansley was not directly involved this time, his name cropped up in many of the letters surrounding the incident. In brief outline, the controversy centred upon a meeting of the British Association for the Advancement of Science (BAAS) planned for Edinburgh in 1921 – the first time the BAAS had visited Scotland in over 20 years – and the announcement that Mrs Agnes Arber (née Robertson) would be President of the Botanical Section (K), a position generally agreed to be prestigious. A group of senior botanists swiftly mobilised to block her appointment. Among their objections, they argued that the normal appointment procedure had not been followed, so that Arber had been appointed without the required wider consultation; she had simply been proposed by the current President, her Cambridge colleague, Miss Rebecca (Becky) Saunders. Not surprisingly, the Scottish professors, Bower and Balfour, were outraged because they had had no say in the planning of a major Scottish meeting. Others, such as Seward, objected to one female president being followed by another, saying a 'botanical gynocracy is unacceptable'.[33] Some thought Cambridge was gaining too much influence, while many thought Arber was simply not worthy of the presidency. Oliver was one of Arber's few supporters but he soon recognised the strength of the opposition to his ex-student and colleague; her swift and dignified withdrawal may well have followed advice she received from him.

Getting rid of Arber was relatively easy. Agreeing whom should replace her was much more difficult. Both Bower and Lang included Tansley's name on their lists of possible replacements but, as long as Balfour remained in a position of influence, Tansley stood little chance of selection. Writing to the overall President of the BAAS, Professor Herdmann of Liverpool, to object to Arber, Balfour had warned against choosing Tansley, whom, he told Herdmann, he regarded as a '*persona irritans*'. In the event, the elderly, safe, D. H. Scott replaced Arber as president of Section K.

The depth of Balfour's dislike of Tansley cannot be overestimated. We know nothing of the personal chemistry between the two men, but there are some pointers as to why, professionally, Balfour found Tansley so irritating. It has already been mentioned how Tansley had read and appreciated the significance of Warming's *Lehrbuch* long before Balfour and Groom had finished translating the book into English. While they were labouring, Tansley was organising and mobilising, positioning himself to become the managing director of ecology. Balfour may also have thought that Tansley had snubbed him when he launched the *New Phytologist*. Fifteen years earlier, in 1887, Balfour himself had, with the help of Sydney Vines, launched another British journal, the *Annals of Botany*. In the words of Bower, a leading supporter of the *Annals*, its foundation was 'a landmark in the British revival [of botany] … Not only did it provide a permanent home for memoirs which hitherto were without one of their own … it also serves to keep its subscribers informed as to the latest movements of British research'.[34] In his history of the period, Reynolds Green (1914) judged that Balfour had become accustomed to 'imposing his personality on the literature of the time'.[35] The upstart *New Phytologist* might have been seen to undermine not only the *Annals*' position but also Balfour's. While Keeble was to be excused his part in the Manifesto affair, Tansley had offended Balfour in too many ways to be forgiven.

Fortunately for Tansley, Balfour was very close to the end of his career. His health was poor – a decline that had been hastened by the loss of a son in the war – so by the early 1920s he rarely left the Royal Botanic Garden in Edinburgh, preferring to express his trenchant views via letters.[36] After a short retirement, Balfour died at the age of 69 in the autumn of 1922. It seems more than a coincidence that Tansley was chosen to be President of Section K of the BAAS meeting in 1923.

***

When Ernest Jones, Britain's leading psychologist, learned that Tansley was about to visit Sigmund Freud in Vienna, a meeting he had helped to arrange, he wrote to Freud, 'Has Tansley started yet? I think he is a very able and careful thinker, and I shall be glad to hear your impressions of him'. To which Freud replied, 'Tansley has started analysis last Saturday. I find a charming man in him, a nice type of English scientist. It might be a gain to win him over to our science at the loss of botany'.[37] It was to be a close run race, but in the end it was botany that won Tansley's lasting attention.

If he was not lost to botany, why is his involvement with psychology of importance? The answer is twofold. Firstly, it reveals a lot about the man himself, about the way in which, risking scandal, he allowed himself to fall in love with one of his own students and, also, about the way in which he saw his own academic prowess in relation to that of the acknowledged leader of another discipline. Secondly, what Tansley learned about psychology informed his ecological thinking.

The origins of Tansley's interest in psychology are not clear. His sister's condition may have awakened his interest, or possibly it stemmed from the early 1890s and those late-night discussions with Russell and other friends during their undergraduate days at Trinity College. What is known is that in 1908 one of his ex-students from UCL, Dr Bernhard Hart, who was working as a doctor at Longrove Asylum, Epsom, attracted favourable attention when he published an article, 'A philosophy of psychiatry'.[38] Encouraged, he followed this with what came to be accepted as the standard textbook on mental disorders, *The Psychology of Insanity* (1912).[39] Around this time, and certainly beginning before he moved to Cambridge, Tansley began visiting Hart, witnessing examples of insanity at first hand and, no doubt, discussing their theoretical background with the man who was at the cutting edge of the study of extreme mental disturbance.[40] A Cambridge botany student, E. Pickworth Farrow, recalled that Tansley brought proofs of Hart's book into the classroom.[41] Friends, and fellow students of botany, Lucy Wills (who completed her studies in 1911) and Margot Hume, both left Newnham College, Cambridge with a strong interest in Freud,[42] most probably encouraged, if not initiated, by Tansley. And Joseph Needham, who was to distinguish himself as a biochemist and sinologist, experienced Tansley talking about Freud during lectures in the early 1920s.[43]

In 1915 Tansley had a dream that was still fresh and vivid when he described it to Kurt Eissler in 1953, as the latter was compiling the Sigmund Freud Archives:

> I was in a sub-tropical country, separated from my friends, standing alone in a small shack or shed which was open on one side so that I looked out on a wide

open space surrounded by bush or scrub. In the edge of the bush I could see a number of savages armed with spears … they showed no sign of hostility. I myself had a loaded rifle, but realised that I was quite unable to escape in the face of the number of armed savages who blocked the way.

Then my wife appeared in the open space, dressed entirely in white, and advanced towards me quite unhindered by the savages, of whom she seemed unaware. Before she reached me, the dream, which up then had been singularly clear and vivid, became confused, and though there was some suggestion that I fired the rifle, but with no knowledge of who or what I fired at, I awoke.[44]

The dream was told to Eissler in order to explain why Tansley decided to read Freud's works in their original German language, something he was well able to do thanks to the German he had taught himself while an assistant at UCL. 'My interest in the whole subject was now thoroughly aroused, and after a good deal of thought I determined to write my own picture of it as it shaped itself in my mind'.[45]

In their detailed analysis of the dream, Cameron and Forrester (1999) present two interpretations. In the first, or Tansley's version, the dream was set in South Africa because that was where a number of his old pupils (the friends) had gone, including the girl with whom he admitted he had fallen in love. The tension was between his wife, whose white clothing and unawareness of the threat that surrounded her symbolised her sexual purity, and his 'beloved'. The savages were the 'herd' – not people of his world, among whom his reputation was secure – but inferior beings whose opinions would be damaging if his infidelity became public knowledge. An alternative, or further layer of, explanation, is that his wife represented botany (what Freud was to call Tansley's 'mother subject'), while his new love represented psychology.

Whatever its interpretation, his dream led Tansley not only to study Freud's work more intensively and to interpret his own dream but – with a little help from Hart who criticised the manuscript – to write his own account of the workings of the normal human mind. His book, *The New Psychology and Its Relation to Life*, was published in June 1920, becoming an instant hit, and reprinted twice within eight months, selling over 10 000 copies in the UK within the first three years. A revised and enlarged edition was published in 1922 and the book was translated into German and Swedish.

Among the overwhelmingly favourable comments on the first edition, Ernest Jones wrote in the *International Journal of Psychoanalysis*, 'It would be difficult to rival it as an introduction either to psychology in general or to clinical psychology in particular … Though written for the educated public at large, it could be read with much profit by any medical man, sociologist, or anyone who desires to be informed as to what is vital in present-day psychology'.[46] As pleasing as was the overall response, the publicity it attracted brought problems for it encouraged many urgent pleas for practical assistance from people who were unaware of Tansley's lack of professional qualifications. It was at this point that Tansley, realising that he needed help, asked Ernest Jones to give him an introduction to Freud. Arrangements were made for him to spend three months in Vienna, from March to June 1922.

Feeling that this short period had been inadequate, and pulled ever more strongly away from botany and towards psychology, a second visit was arranged. Tansley resigned his Cambridge lectureship in spring 1923, on 8 March writing to Clements:

> Probably I shall cease to be a professional botanist after the [university] term, though for the present at least, I shall continue to edit two journals ... Adamson is going to the Cape [leaving Manchester] and will be a terrible loss to me – I need a good 'florist' at my elbow. Together with the 'conservatives in authority' his departure will help to make me spend more time at psychology and less at ecology.[47]

While finalising arrangements for his second visit, Tansley kept in touch with botany. He fulfilled his duties as President of Section K of the BAAS meeting in Liverpool, although, as Godwin noted, 'a curious aura of loneliness attended him'.[48] Together with Godwin, he passed much of the summer of 1923 continuing their research at Wicken Fen and, no doubt, since Godwin too was a keen student of Freud, discussing psychoanalysis and Tansley's next visit to Vienna.

For this second visit, Tansley took Edith and their children with him. Arriving in Vienna in October, the family was immediately swept into Vienna's busy social whirl. Tansley's analysis was unavoidably delayed until late December because Freud was undergoing a series of operations for oral cancer, so the Tansleys were free to enjoy the pleasures of alpine walking and, with the onset of winter, skiing. Tansley's second period of analysis was lengthier than the first. Few details are known but it triggered a series of letters from Arthur to his sister Maud in which he tried to find out more about the formative years of his own childhood, such as the age at which his cot was taken out of his parents' bedroom. It was six months before the Tansleys returned to Grantchester.

In later life, Tansley recalled that he had been disappointed by his sessions with Freud; no great discoveries of forgotten scenes from his childhood had been revealed and Freud spent more time discussing theoretical questions than Tansley's own subconscious. The cause of the disappointment may, however, have lain with Tansley himself rather than Freud. As Forrester and Cameron have pointed out, Tansley had already interpreted his own dream with 'complete conviction';[49] he believed he was not neurotic and did not need analysis. He liked to think of himself as Freud's equal, once describing their relationship as akin to that of 'two sovereigns' discussing analysis.[50] For his part, Freud once told Ernest Jones that in analysis 'Tansley is bringing up enormous resistance'.[51] In spite of such difficulties, mutual admiration grew between the two men and they exchanged a series of gifts as tokens of a genuine friendship (Chapter 9). When Freud died in 1939 the obituary that Tansley wrote for the Royal Society was, in Godwin's words, 'a movingly sincere and exact tribute'.[52]

Somewhat surprisingly, given the warmth with which he was welcomed into the psychoanalytical community – in the summer of 1924 he was made an associate member of the British Psycho-Analytical Society, and in autumn 1925 a full member – Tansley's involvement stuttered after his return to Cambridge. As Freud had recommended, he took on a case, 'an experimental analysis, lasting nearly two years, on an obsessional neurotic',[53] he attended conferences at home and abroad, and he vigorously defended psychoanalysis in the press,[54] but there was never total commitment. Whether he was held back by an awareness of his lack of medical qualifications, as he sought to establish himself in a young, self-conscious profession where such qualifications were increasingly expected as the norm, or by the continuing pull of botany to which he still had a day-to-day commitment, is not known. What is known is that a psychoanalysis discussion group coalesced around him in Cambridge.[55] It was composed of younger men, two of whom, Lionel Penrose and Harold Jeffreys, were to make not only important contributions to the young and developing subject of psychoanalysis but who also had interests in ecology – they published, respectively, in Tansley's *New Phytologist* and *Journal of Ecology*.[56] (The contributions of a third member, E. Pickworth Farrow, to both ecology and self-analysis are detailed in Chapter 12.) Tansley continued to be passionately interested in the development of psychoanalysis, talking and writing about it until the end of his life, but when, late in 1926, the Sherardian Chair of Botany at Oxford once again became available the clouds of confusion at last cleared from his mind. As Freud had perceptively forecast, Tansley would return 'to the mother subject'.[57]

The second reason why Tansley's involvement with psychology was important is that, as mentioned earlier, it affected his view of vegetational processes. At the core of his new psychology was his belief that, 'the abnormal activities of the mind, as seen in cases of hysteria and insanity, are but extreme and unbalanced developments of characteristics and functions which form integral parts of the normal healthy mind'.[58] When he and Godwin began their studies of Wicken Fen, he soon realised that the same principles were applicable to the dynamics of vegetation.

Harry Godwin was a brilliant product of the Cambridge Botany School who had first made his mark when still an undergraduate. He had published a short paper showing that the floristic richness of a number of local ponds was correlated with their age.[59] His interest in ecology had been kindled by Tansley's lectures but his postgraduate research was done under F. F. Blackman's supervision.[60] At interview, the young man had told Blackman that his purpose was 'ultimately to apply the methods of plant physiology to ecology'. To which Blackman replied tartly that he had found plant processes hard enough to explain when he had the subjects in the laboratory under controlled conditions, without examination in the wild, where every factor of the environment might vary independently. However, Blackman not only took on Godwin but tolerated his exchanging the confines of the laboratory one day each week for the fresh air and his plots of vegetation at Wicken Fen.[61]

Part of Wicken Fen had been owned by the National Trust since 1899 (p. 5), having been selected because it was, according to the Trust's Annual Report, 'almost the last remnant of primaeval fenland', shaped by centuries of sedge-cutting and peat-digging.[62] The Fen, which is today a National Nature Reserve encompassing more than 700 acres, includes sedge beds, reed communities, and meadows, as well as drier woodland and grasslands. It lies at the southwest corner of a much larger area, stretching almost from Cambridge to the east coast, which was wetlands until drained by Roman and Dutch engineers. After Tansley had founded the Cambridge Ecology Club in 1921, he, Godwin, and other founder members made an inaugural excursion by bicycle to Wicken Fen where they recognised a ready-made open air laboratory, a place where experiments could be carried out without 'interference ... or destruction of the vegetation of the area under observation'.[63] With plans laid, permissions obtained, and financial assistance from Tansley, Godwin was in 1923 able to fence off a small plot for his studies, an area that became known as the 'Godwin Triangle'. Two more areas were set aside in 1927, again made possible thanks to Tansley's financial support.

Vegetation within the 'Triangle' was not cut or managed in any way, its natural succession being slowly revealed to Godwin as year by year he mapped the plants. Within the additional plots, sections were cut at intervals of one, two, three, or four years, or left uncut. Thus, natural succession was 'deflected but not arrested' in the cut plots.[64] The different intervals between cutting revealed the extent to which different species were suppressed by cutting. Tansley called such deflected successions 'plagioseres' (where a 'sere' is an intermediate stage in succession); if the deflecting factors were removed then vegetation reverted to the normal development of water-logged ground, the 'hydrosere' (Figure 8.1). The parallels with the human mind, deflected from its normal healthy development into insanity or hysteria, are clear.

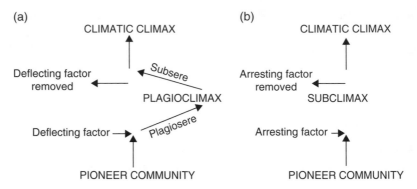

Figure 8.1   (a) Deflected succession and the plagioclimax. (b) The subclimax, as conceived by Frederic Clements. (After Tansley 1939a.)

The long-term studies of Godwin and Tansley at Wicken Fen had practical and theoretical implications of the greatest importance. Among the former, it was now clear that without active management:

> The mixed sedge and litter will all become quickly colonised by bushes, and as the Fen passes by rapid stages to carr most of the characteristic fen flora and fauna will disappear. A rigid preservation of the Fen 'in its natural state', with no cutting of vegetation or clearing of lodes [main waterways] and drains, is the quickest way to exterminate most of the species it is desired to conserve.[65]

Here was hard evidence in favour of conservation over preservation, a demonstration of the need for nature reserves to be studied and then managed by experts with detailed local knowledge.

***

Excluding those such as Balfour whose dislike of Tansley was of a personal nature, there was some feeling within the botanical community that Tansley was a theoretician, a talker rather than a doer:

> [Tansley] is a rather dreamy philosopher, who curiously enough combines that with a singular esteem of his own opinions, which others do not always share.
> *F. O. Bower to Sir William Bate Hardy, 26 October 1921*[66]

If there was some justification for such a potentially damaging impression then, in the nine years between his first, unsuccessful, application for the Oxford chair and his second, successful, application, Tansley dispelled it. His paper on competition between *Galium* species (1917) had proved he could employ small-scale experiments to support his theories.[67] From a practical study of durmast oakwoods (*Quercetum sessiflorae*) in the Malvern hills,[68] conducted with his ex-UCL colleague, Edward Salisbury (by then a lecturer at East London College and fast becoming an authority on successional series), Tansley had been able to reconstruct the vegetational history of the Malvern hills, an area he had known and loved since childhood holidays. The local limestones were once dominated by ash, with an accompanying limestone flora, the authors concluded. However, in the more southwesterly Malverns the acid-loving (calcifuge) durmast oak had moved in as leaching had taken effect and humus had accumulated, restricting ash to the steeper slopes where humus could not accumulate and fresh chalk was regularly exposed. The general point made by the authors was that calcareous rocks do not always give rise to calcareous soils; seedlings of durmast oak were thus able to establish themselves in a region of Silurian limestone bedrock because they grew in a surface layer of acidic humus during their critical early stages of development.

Ever since 1909, when he set up his first rabbit-proof enclosures at Ditcham Park in Hampshire, Tansley had been monitoring the effect of rabbit grazing on chalk grassland. The botanical community was largely unaware of these long-term studies until his results began to appear in the *Journal of Ecology* in 1922.[69] There followed in 1925 and 1926 two further papers in which he described joint studies made with Adamson along a 50-mile belt stretching from Butser Hill in the west (eight miles north of Kingley Vale) to Chanctonbury Ring in the east and onward to the Channel coast. Starting in 1921 by making detailed lists of the occurrence and abundance of plant species, and analysing soil pH and nutrient content, the authors concluded that 'chalk grassland is a phase in the development of vegetation on chalk soil, and that the mass of it is a biotically determined climax association, its characteristics being fixed by the continuous grazing factor' (cattle and sheep as well as rabbits).[70] Put simply, much if not the whole of the chalk grassland would pass into forest if pasturing and rabbit attack were altogether stopped.[71] The authors noted that, in Clements' terminology, 'it [chalk grassland] is of course a *subclimax* – the true climatic climax being beechwood'. If grazing intensity were diminished, flat surfaces and gentle slopes became dominated by *Erica* and/or *Calluna* species (heath), while scrub leading to ashwood and beechwood prevailed elsewhere.

The results of Tansley's other long-term project, at Wicken Fen, were not published until 1929, after he was in post at Oxford. However, there is no doubt that by the time of his second application the botanical community had been made well aware of Tansley's credentials as a practical scientist.

This period cannot be left without reference to Tansley's publications. He wrote two small books, *Elements of Plant Biology* (1922a), for medical students, and *Practical Plant Ecology* (1923), for schoolchildren, to which we will return in the final chapter. However, it was his third publication that had immediate significance for it helped him at last to win the Oxford chair.

In the summer of 1924 the Imperial Botanical Conference, held in London, appointed a committee to provide ecological guidance to foresters, agriculturalists, and entomologists working in the mainly tropical and sub-tropical countries of the Empire. The British Empire Vegetation Committee, as it was called, was chaired by Tansley, and on it he was joined by two ex-colleagues, Frank Oliver and Edward Salisbury, and also by John Ramsbottom (Keeper of Botany at the Natural History Museum). The Secretary to the Committee was Thomas F. Chipp, a forester with extensive experience in Malaysia and West Africa, who had been since 1922 Assistant Director at Kew. The tangible outcome of the Committee's work was a large book, *Aims and Methods in the Study of Vegetation* (1926). Tansley and Chipp were the editors of this multi-authored volume, and they wrote most chapters in Part I, including 'Nature and aims of the study of vegetation', 'Plant communities', 'Factors of the habitat', 'Training', 'Methods of investigation', and 'Collecting'. Part II concentrated on the vegetation of specific parts of the Empire. Always an astute businessman, Tansley arranged for the book to be distributed by the Crown Agents, thus avoiding booksellers' charges: the intended profits were to be used to support the publication of a series of monographs on the vegetation of different parts of the Empire. Sadly, fire destroyed the first stock and any profits that might have been realised.[72] Only one monograph ever appeared, Adamson's *Vegetation of South Africa*.[73]

Forestry was one of the main themes running through *Aims and Methods*. Tansley had always had a special interest in forestry since, he believed, in order for the forester to be effective, a knowledge of ecology was essential. If the resource that vegetation represented was to be managed efficiently, its properties had first to be fully understood. The significance of this book for Tansley's career was that, fortuitously, one of the contributors to Part II was Robert Troup, and Troup was to be one of the electors for the Oxford chair in 1927 (Chapter 9).

***

Tansley never lost his interest in psychoanalysis. Deep into retirement, at the age of 80, he was in 1951 among a small group of distinguished men who backed an appeal by the Institute of Psycho-Analysis to raise £100 000 to fund its new London headquarters, teaching, and clinical work.[74] The next year he published his last book, *Mind and Life*. As in *The New Psychology* (1920), he argued that the balance of, physically unmeasurable, psychic energies was the key to mental health. He proposed that the healthy mind tends towards equilibrium, or stability, in the face of an array of external stimuli.

Tansley's last meeting with Sigmund Freud had occurred many years earlier, in the summer of 1928. He had travelled to Vienna from an International Phytogeographic Excursion (IPE) based in nearby Czechoslovakia and Poland. As was the custom of the day, Tansley's entry to Freud's study was preceded by the presentation of his visiting card, on which he announced that he was Sherardian Professor of Botany at the University of Oxford. Freud's response as he greeted his visitor was forever etched in Tansley's memory,

> he immediately enquired 'Ordentlich [full professor]?' 'Jawohl', I replied. 'Das is gut', he said. He had had enough experience of being 'aussererordentlich [associate professor]' himself to be acutely conscious of the difference, and was unfeignedly glad of my new academic status, which he was sure would be good for me psychologically.[75]

The approval of Freud was a huge boost to Tansley's self-esteem for, although he may have thought himself Freud's equal, he once confessed to Oxford colleagues that he thought Freud was 'the most important thinker since Christ'.[76] Tansley's use of the emphatic 'Jawohl', rather than a simple 'Ja', reveals his pride in his new status. His professorship, finally attained in 1927 at the age of 56, was both a relief and a release. As Freud so correctly judged, it would be good for Tansley psychologically. It completed his return to botany after a long dalliance with psychoanalysis. It helped heal the wounds he had suffered in defending ecology.

Similarly, for his wife and family, time soothed the pain he had caused, although an unhappiness remained with them always.

Notes

1. Godwin 1977, p.12.
2. Anon. 1921a. Abandonment of summer excursions. *Journal of Ecology* **9**, 106. In 1920, excursions to Blakeney Point and Esthwaite (Lake District) were attended by 11 and 20 members, respectively (Anon. 1921b. Annual meeting. *Journal of Ecology* **9**, 100).
3. Tansley Archives, University of Cambridge Library.
4. Boney 1991, p.14.
5. Cameron, Forrester 1999, p.67.
6. Tansley Archives, University of Cambridge Library.
7. Ibid.
8. Godwin 1977, p.13.
9. Sheail 1987, p.30.

10. Ibid., p.37.
11. Allen 1986, p.102.
12. Tansley 1916, p.82.
13. Marsh AS. 1915. The maritime ecology of Holme next the Sea, Norfolk. *Journal of Ecology* **3**, 65–93.
14. Marsh AS. 1914. Notes of the anatomy of *Stangeria paradoxa*. *New Phytologist* **13**, 18–30.
15. Walters 1981, p.92.
16. Tansley AG. 1917. On competition between *Galium saxatile* L. (*C. hercynicum* Weig.) and *Galium sylvestre* Poll (*G. asperum* Schreb.) on different types of soil. *Journal of Ecology* **5**, 173–179.
17. Tansley 1916, p.81.
18. Ibid., p.85.
19. Entering Cambridge in 1919, Harry Godwin mixed with a generation who had served in the trenches. 'Their eyes were haunted, but they all displayed a kindness to one another and a gentleness to the schoolboy freshmen that were all-embracing: it was enough that you and they were still living for them to offer a friendship embracing everything they were possessed of' (Godwin 1985b, p.18).
20. H29 Tansley Archives, University of Cambridge Library.
21. Godwin 1977, p.13.
22. Royal Botanic Gardens, Kew, 'Miscellaneous correspondence': Tansley to Cotton, 14 September 1915.
23. Russell was deprived of his Trinity Fellowship and imprisoned in 1916 for his protests against the war.
24. Bower 1918, p.107.
25. Blackman *et al.* 1917, p.242.
26. Boney 1991, p.6.
27. Ibid., p.10.
28. Ibid., p.18.
29. Golley 1993, p.208.
30. Dukinfield Henry Scott, palaeobotanist. In 1882, at age 28, he had been appointed Daniel Oliver's assistant at UCL. He helped teach Frank Oliver and the two became lifelong friends. He was later Keeper of the Jodrell Laboratory at Kew (Chapter 6).
31. Morrell 1997, p.235.
32. Boney 1991, p.17.
33. Boney 1995, p.30.
34. Bower 1938, p.64.
35. Green 1914, p.575.
36. Boney 1995, p.34.
37. Cameron, Forrester 1999, pp.65–66.
38. Hart B. 1908. A philosophy of psychiatry. *Journal of Mental Science* **54**, 473–490.
39. Hart B. 1912. *The Psychology of Insanity*. Cambridge: Cambridge University Press.
40. Cameron, Forrester 1999, p.78.
41. Cameron, Forrester 2000, p.229.
42. Bastian 2008, p.90.

43. Cameron, Forrester 1999, p.74.
44. Ibid., p.65.
45. Ibid., p.69.
46. Tansley 1920, frontispiece.
47. Cameron, Forrester 1999, p.71.
48. Godwin 1977, p.14.
49. Cameron, Forrester 1999, p.82.
50. Ibid., p.83.
51. Cameron, Forrester 1999, p.85.
52. Tansley AG. 1939–41. Sigmund Freud, 1856–1939. *Obituary Notices of Fellows of the Royal Society* **3**, 246–275; Godwin 1957, p.236.
53. Cameron, Forrester 1999, p.73.
54. Tansley AG. 1925a. Psycho-analysis. *The Nation and The Athenaeum*. 8 August, p.567; Tansley AG. 1925b. Freudian psycho-analysis. *The Nation and The Athenaeum*. 12 September, p.700 (cited by Forrester and Cameron 1999, p.74). Tansley pointedly finished the first letter above, 'may I beg your correspondents' attention to the fact that I am not, and never have been, a professor? Nor do I hold a doctor's degree'.
55. Cameron, Forrester 2000, p.192.
56. Ibid., p.201, p.224.
57. Godwin 1957, p.236.
58. Cameron 1999, p.10.
59. Godwin H. 1923. Dispersal of pond floras. *Journal of Ecology* **11**, 160–163.
60. In contrast, Tansley found it difficult to collaborate with F. F. Blackman, who rose at 11 am and worked after dinner until 4 am. Tansley was an early riser, believed in a quiet leisurely dinner with alcohol to aid digestion, and no work later (J. L. Harley 1990, Archives of the New Phytologist Trust).
61. Godwin 1985b, p.57, p.146.
62. Murphy 1987, p.113.
63. Cameron 1999, p.12.
64. Ibid. p.13.
65. Ibid., p15.
66. Glasgow University Archives 0248 DC002/14/220.
67. Tansley 1917 (see note 16).
68. Salisbury, Tansley 1921.
69. Tansley 1922b, pp.168–169.
70. Tansley, Adamson 1926, p.32.
71. Ibid. p.3.
72. Anker 1999, p.37.
73. Godwin 1957, p.237.
74. Anon. 1951. News, notes and comments. *International Journal of Psycho- Analysis.* **32**, 269.
75. Cameron, Forrester 1999, p.86.
76. Godwin 1977, p.25.

9 The Oxford Years, 1927–1937

Unemployment was the dominant social issue in the inter-war years as the immediate post-war economic boom of 1919 soon collapsed. Levels of unemployment in Britain were rarely to dip below 1 million and in 1931 they would reach a peak of 3 million – reckoned at the time to be 20% of the (potential) working population. On the one hand, the nation's heavy industries, located mainly in northern England and Scotland, were shedding labour as manufacturing processes became automated. On the other hand, its more modern, lighter industries, located mainly in the south, were impoverished by World War I (WWI) and had invested little in the latest technical developments. Britain was fast losing overseas markets to its competitors, often in Asia, and there were fears of those cheap Asian goods being 'dumped' on British markets. The country's multiple problems were exacerbated when its great trading partner, the USA, suffered the Wall Street Crash (1929) followed by the Great Depression. Arthur Tansley was for four of those years one of the unemployed, though in his case it was by choice, cushioned as he was by family wealth and still able to live a gentleman's life on his own terms. In accepting the Sherardian Chair in 1927 he exchanged the gentle atmosphere of Grantchester for the quiet seclusion of Magdalen College, Oxford. He was effectively swapping one cocoon for another. Even the ancient city to which he was moving was sheltered from the industrial malaise affecting most of the country for Oxford's major employer, after the University, was motor manufacturing, which was one of the few industries to prosper at the time.

The worlds of ecology and conservation did not escape the economic difficulties of the post-war years for it would have been politically unwise for any government to spend its limited resources on what would have been viewed by most taxpayers as causes of low priority.

Shaping Ecology: The Life of Arthur Tansley, First Edition. Peter Ayres.
© 2012 by John Wiley & Sons, Ltd. Published 2012 by John Wiley & Sons, Ltd.

... in terms of legislation, public awareness or [the acquisition of] nature reserves, the tangible achievements of the inter-war years were meagre.

Although politicians such as Ramsay MacDonald and Neville Chamberlain[1] gave personal support to the preservation and recreational movements, central government consistently refused to set a precedent for the use of Exchequer monies ...

J. Sheail, *Nature Conservation in Britain*[2]

Shortly before the outbreak of WWI, Frank Oliver had reviewed for the British Ecological Society (BES) the main problems facing those who were trying to create reserves in Britain, not least of which was the lack of any perceived threat to the countryside. As he succinctly put it, 'One cannot raise money for a thing that is not threatened'.[3]

The increasing difficulties faced by Oliver and like-minded men in between-wars Britain are illustrated well by what happened in Norfolk, and contrast with the pre-WWI situation when the National Trust's first acquisiton of land at Blakeney had been relatively easy. Recognising the scientific studies that Oliver and his students had been carrying out on that stretch of coast including Blakeney, its owner, Lord Calthorpe, was in 1910 pleased to grant a lease to University College, London (UCL) for further 'marine horticulture', seemingly protecting the land from unwanted development. Nevertheless, when Calthorpe died shortly afterwards his heirs put the land up for sale, with the implicit threat that it might be built upon by a purchaser. However, led by Charles Rothschild, and with major assistance from Oliver and also G. C. Druce (who persuaded the Fishmongers' Company to make a particularly large donation), enough money was raised by private subscription to enable Rothschild to purchase 1110 acres of sand dunes, scrub, shingle, and marsh comprising Blakeney Point and slightly beyond. Ownership of the land was then transferred to the National Trust in the summer of 1912.[4]

In the post-war years such large purchases were rare since not just public but also private finances were being increasingly stretched. Exceptionally, Scolt Head Island, just off the north Norfolk coast, was added to the Trust's lands in 1923, thanks again to the efforts of Oliver and local donors who were able to purchase the island from Lord Leicester. More often, as happened when 400 acres of nearby Cley marshes were bought by public auction in 1926, the Trust refused to take over responsibility for the extra lands. The local organiser at Cley, Sydney H. Long, was forced to establish a non-profit-paying company, called the Norfolk Naturalists' Trust, to purchase and administer the area.[5] The establishment of those few reserves created in the between-wars years, and of many since, followed a similar pattern which involved the rallying of strong local support for the protection of a relatively small area.

On top of economic problems, the preservation movement in general, and the Society for the Promotion of Nature Reserves (SPNR) in particular, suffered a terrible blow in 1923 when its crusading leader, Charles Rothschild, fell victim to the Spanish influenza epidemic that was killing millions across Europe. The departure of Oliver, when he moved to Egypt after retiring from UCL in 1929, was yet another setback.

Before leaving this period between the wars, one other event is worthy of mention because it illustrates the changing mood of the times – the Kinder Scout Mass Trespass of 24 April 1932. Kinder Scout, 631 metres above sea level, lies at the heart of the gritstone region of the Peak District, whose vegetation had been extensively surveyed by Charles Moss (Chapter 7) and Robert Adamson (1918). An area of poor moorland, about 60 700 hectares in extent, it was devoted either to sheep grazing or the rearing of game birds for shooting, with the public was excluded from all but 485 hectares. There were only 12 legal paths across these private lands. Rambling had become very popular by the late 1920s because even for the unemployed masses – or especially for them – it was a free and healthy way of lifting the spirits. It is estimated that up to 15 000 ramblers left Manchester every Sunday, many heading for the Peak District. Responding to the huge public pressure that had built up for greater access to the countryside, a group of 400 ramblers coming from the Manchester direction and 200 coming from the Sheffield direction, deliberately trespassed on the private lands and met on the plateau of Kinder Scout. Six of the leaders were subsequently arrested, convicted, and imprisoned for between two and six months, but their cause and their treatment unleashed a huge wave of public sympathy.[6] Nature reserves and conservation were not the interest of these law breakers or their sympathisers, but the Mass Trespass was symptomatic of growing public pressure to open up the wild places of Britain. And that pressure would lead before long to the creation of national parks in a political process that was both aspirationally and practically entwined with the creation of the Nature Conservancy and National Nature Reserves (Chapter 11).

***

An academic appointment can depend as much upon the composition of the appointing committee as upon a candidate's curriculum vitae. Tansley's rejection by Oxford in 1919 happened in large part because the electors included two men, D. H. Scott and I. B. Balfour, who already had a low opinion of him. They, and A. C. Seward, were the botanists in the group and, presumably, their views weighed most heavily. The composition of the appointing committee in 1927 was completely different, and this time it favoured Tansley.[7] The electors included Edwin Goodrich, Linacre Professor of Zoology in Oxford, Robert Scott Troup, Professor of Forestry at Oxford's Imperial Forestry Institute, and John Bretland Farmer, Professor at Imperial College, London. Goodrich had in his younger days worked for a period as assistant to E. Ray Lankester, so it might be expected that he was already sympathetic to Tansley's philosophy. This time Troup and Farmer were the botanists. Working strongly in Tansley's favour was the fact that Troup had only recently contributed a chapter, 'Problems of forest ecology in India', to Tansley and Chipp's *Aims and Methods in the Study of Vegetation* (1926) (Chapter 8). Before taking the chair in Oxford, Troup had worked in many countries of the Empire and was a strong supporter of the aims of the British Empire Vegetation Committee; as such, he would have been well aware of Tansley's interest in forestry at both home and abroad. If Troup and Tansley were kindred spirits, then Farmer and Tansley were not. Farmer was a protégé of Balfour, once being described as Balfour's 'star pupil'. He had naturally sided with the Morphologists in the Tansley Manifesto affair (Chapter 8) and was a potential threat to Tansley's selection. If open opposition was expressed by Farmer, who was noted for his forceful and dominating character,[8] then it must have been overridden by the other electors because Tansley was offered the Sherardian Chair.

The post carried with it a Fellowship of Magdalen College (Figure 9.1). Founded in 1588, Magdalen is one of the University's oldest, largest, and architecturally most attractive colleges. Its hall, chapel, cloisters, quadrangles, and New Building (1733) lie on the west bank of the river Cherwell, while its extensive meadows and deer park are on the east bank. The setting is idyllic. Facing Magdalen, separated from it only by the High Street, is the gateway to the University's Botanic Garden. To the immediate west of the gateway is the

Figure 9.1   Magdalen College from High Street, Oxford, in 1925, photographed by Tansley's colleague, A. H. Church. The Department of Botany (hidden, right) is opposite the tower. (© Images and Voices, Oxfordshire County Council, with permission.)

Daubeny building, where in 1927 was located the Department of Botany, sharing space with Chemistry. Built in 1848 by Charles Daubeny, who was Professor of Chemistry, and of rural economy, as well as of botany – and largely financed from his own pocket – the exterior elegance of the classically styled building hid an interior ill suited to the teaching of sciences in the 20th century.[9] The lack of suitable teaching, research, and office space proved a constraint from which Tansley never managed completely to free himself or his staff.

By tradition, newly appointed professors give an inaugural lecture to introduce themselves to the University. They may choose to explain their subject, if it is a new one, outline their own work, or, as in Tansley's case, set out a vision for the future of their department. In his lecture given on 22 November 1927, the *utility* of botany was the starting point for the 55-year-old Tansley. To what was a theme familiar to his older friends and colleagues, that a study of pure botany underpins both forestry and agriculture, he added a new dimension, 'the natural pastures of the world'. 'The scientific study of grasslands is the necessary foundation', he said, 'of their proper economic treatment as the grazing lands of the world'.[10] And where were most of the those grasslands found? In the British Empire. And where were the most numerous opportunities for the employment of botany graduates? In the British Empire. None too subtly, he presented a case, clothed in patriotism, for expansion of the Botany Department. Before an audience which would have included the most senior and powerful men in the University, he floated the idea that the department might provide basic courses for students in agriculture, forestry, and in rural economy, while reminding them that botany's current accommodation was far too small to cope with significantly larger classes. Whether or not botany should move from the Botanic Garden to what was called the 'Museum Site' (a mile away and now at the heart of the science area in South Parks Road) was a major issue throughout Tansley's years in Oxford. It was a problem that neither he nor the University solved satisfactorily.

When measured against his predecessor in the Sherardian Chair, Frederick Keeble, Tansley had little to live up to. Seward had been right (Chapter 8), Tansley would have been much the better choice in 1919.

> Keeble was not only at odds with his demonstrator, A. H. Church, but as a horticulturalist [he] opposed the application of mathematics to botany, the recently inaugurated DPhil, specialisation and the use of experiment, measurement and physical methods in general in the study of plant life.[11]

Keeble's interests in Oxford were the four 'P's, planting, pruning, potting, and propagating, plus Lillah McCarthy, an actress whom he married in 1920 and with whom he thereafter 'indulged in extensive hospitality'.[12] To be fair to

Keeble, Church was difficult to work with – unsociably aloof, interested in only the very best undergraduates, and unwilling to publish his research because he objected to its being edited by others. Also, Keeble did manage to acquire some extra laboratory space for botany when chemistry moved out of the Daubeny building, together with some research rooms created from vacant museum space. Overall, however, he left for Tansley a small dysfunctional department.

Tansley inherited three departmental lecturers, Church, Henry Baker, and William H. Wilkins. Wilkins had graduated at Oxford in 1921, as a mature student aged 35, but after six years of teaching he had still not developed his own line of research. Tansley moved swiftly to remedy this weakness, suggesting to Wilkins that he should interest himself in fungi (mycology) and arranging for him terms of study under Vernon Blackman's tutelage at Imperial College in 1927–1928. The stratagem worked for, on his return, Wilkins built a successful research career in the study of plant diseases caused by fungi, although he always proved a difficult colleague– possibly bearing a grudge stemming from Tansley's intervention – and was generally resentful of anyone with Cambridge connections.[13] This was unfortunate for Wilkins because, to plug gaps in the department's coverage of botany, Tansley was soon allowed by the University to bring in two young botanists, both from Cambridge. Each was strongly recommended by their PhD supervisor, Tansley's brother-in-law, F. F. Blackman. Will James was a plant physiologist who was to become a world authority on the biochemistry of respiration, while Roy Clapham (see Figure 1.3) was an all-rounder, able to teach anything from taxonomy to genetics and cytology. The number of courses offered each term doubled, attracting more students. The number of finalists in botany had rarely reached double figures in the years leading up to 1927[14] but by the end of Tansley's tenure it regularly exceeded that figure, and the number of postgraduates had quadrupled to 13.

By 1930 Tansley had become so frustrated about the shortage of teaching and research space that, with the knowledge of the University, he approached the independent Pilgrim Trust for £100 000 to pay for a new building, equipment, and an experimental garden in the science area, pointing out that botany at Oxford had only 7000 sq. ft of usable space compared with Cambridge's 23 000 sq. ft, for fewer honours students.[15] Not only was his application refused by the Trust but his hurt was made worse when the University Registrar, on learning of the failure, unhelpfully suggested that the number of botany undergraduates might actually be reduced[16]; this in spite of the University already having an agreement with government to train more biologists. The acute problem of lack of space was ameliorated only when botany gained further sections of the Daubeny building so that there was specialist accommodation for physiology (1932) and mycology (1933), but, as Clapham told his wife, the department's finances were extremely tight:

> The Prof. is having a most worrying time. The Dept. is £90 in debt, receives no money until the end of the quarter, and, by a recent statute, is not allowed to have an overdraft. The plant chambers which we thought would cost £5 have actually cost £14, and Wilkins has been very extravagant lately.
>
> *A. R. Clapham to his wife, 19 January 1932*[17]

Even (junior) staff were called upon to help prepare the new rooms:

> I have been working as an indoor decorator all today, doing some painting in the new laboratory. … Painting is a soothing occupation, but rather messy for the inexpert. Tomorrow I shall finish a cupboard which I started today by giving it an undercoat of paint on the side and a water-stain on the top. There will be a second coat of paint, and the top must be treated twice with linseed oil and once with Ronuk [wax polish]. All this experience should be very valuable eventually!
>
> *A. R. Clapham to his wife, 10 August 1932*[18]

No detail escaped Tansley's attention. In Oxford's Bodleian Library there are thick files full of invoices from local tradesmen for work done in the botany building, each authorised by Tansley. He took a close personal interest in both teaching and research, as is shown by one of Clapham's letters, written just before the start of the academic year 1932–1933.

> This morning he looked at the proofs of the Prelim. exam. papers and at the [anatomical] sections I cut for the exam; also at other sections of Sycamore seedlings for next term, and at the arrangement of benches in the Morphology Lab. That took the whole morning. This afternoon we went to Bagley Wood with Baker and had a careful look at the work he (B.) is doing there, staying there, through storms and fair intervals, until 5.30 pm. I, fortunately, changed my shoes and put an old pair of grey bags [trousers] over my grey suit bags, but the Prof. didn't, and was a good deal muddied and damped in consequence.
>
> *A. R. Clapham to his wife, 27 September 1932*[19]

After several years of Tansley's prompting, guiding, and cajoling, his staff began to attract research grants from the government's Department of Scientific and Industrial Research. To raise further the profile of the department, he encouraged his staff to become 'more visible' through public work, such as sitting on scientific councils and editing journals. Lightening his workload in 1932, Tansley handed over his *New Phytologist* to a group of younger editors, James, Clapham, and Godwin. Together with the journal's ownership, he transferred to them £150 of working capital, taken from the *New Phytologist's* accounts but, in reality, money from his own pocket.[20] He continued to edit the *Journal of Ecology* until his retirement in 1937.

At the time of his Inaugural Lecture, Tansley had had a clear vision of the new Department of Botany he wanted, but he was soon forced to scale down his ambitions as it became obvious that the large sums of money he needed would not be forthcoming from the University. Instead he was forced to pursue much smaller amounts, in the order of £12 000, which would be just enough for modest extensions to the buildings at the Botanic Garden, plus some refurbishment of existing buildings. By 1935 he was still complaining to the University authorities,

> there is scarcely a corner of my own laboratory where an ecologist can work. Two of our men doing post-graduate research in ecology have to work partly in an unheated and unlighted room in the Herbarium.[21]

Thanks to Tansley's unceasing efforts, and with a good degree of self-help (as already seen), accommodation at the Botanic Garden was improved little by little, but there were no new buildings and in his turn Tansley left his successor a miserable legacy. Always alert to his own shortcomings, this failure would have been all the more painful to Tansley because he had been unable to take advantage of 'the Druce bequest'.

***

It is said that Druce disliked Tansley[22] and, although there is no direct evidence, from everything we know about Druce, it is almost certain that Tansley disliked him, or, rather, what Druce stood for. Druce was an amateur botanist whose knowledge of the identity and distribution of plants Tansley had been pleased to utilise when organising the first International Phytogeographic Excursion (Chapter 7). He was the Secretary of the Botanical Exchange Club (BEC; called from 1948 the Botanical Society of the British Isles) from 1903 – two years before he sold his lucrative pharmacy in Oxford's High Street – until his death in 1932. The BEC was the principal forum for amateur botanists in Britain, a club whose *raison d'être* was the sighting and logging of species, but whose aims included the collection, preservation, and exchange of plants. Many members sought to build their own herbarium but none did this more extensively and ruthlessly than Druce.

While he actively supported the SPNR, advised that body on sites most worth protecting, helped it to raise funds to acquire such sites (p. 123), left money to the Society in his will, and generally encouraged 'conservationists' within the BEC – if only to hold his society together – Druce was nevertheless a rapacious collector. As David Allen concluded in his history of the BEC, one explanation of the contradiction in Druce's character is that 'he lived under the delusion that the power which causes a plant to occur in any given spot in the first place can be depended upon to preserve it there in perpetuity'.[23] Conservationists within the BEC soon learned not to show rare specimens to Druce. Allen relates an incident in which Druce was taken to see some rare dactylorchids in Cambridgeshire, whereupon he proceeded to pull them up in armfuls: his guide admitted he would have hit Druce had not another member of the party forcibly restrained him. In his Secretary's Report for 1915, Druce wrote, 'It is a matter of satisfaction to know that all our members are impressed with the importance of avoiding reckless or careless gathering of plants which endanger any local species', yet he included in the same report a member's request for the 'seeds or roots of rare British Specimens'. He was unabashed in his support of the 'collectors' within the BEC.

'King Druce', as he became known, gave unstintingly both his time and his money to the BEC, although his authoritarian control and lack of consultation made him as many enemies as it did friends.[24] Charles Moss was among those who tangled with Druce, although even this combative Yorkshireman eventually withdrew from the battlefield defeated. As he left Cambridge for South Africa, Moss complained of Druce's 'fiendish vindictiveness' (the cause of their dispute is now obscure but it would have been well known to Moss's friend and colleague, Tansley).

In spite of Druce's attitude towards those who opposed him, he was desperate to ingratiate himself with the rich, especially the titled rich, and with the University of Oxford. However, to his frustration, he was never able to penetrate beyond the fringes of the University, an establishment to which he was sometimes useful, and sometimes not. Oxford awarded him an Honorary MA on the strength of his *Flora of Oxfordshire* (1886), in which he described a grass new to science, *Bromus interruptus*. He was encouraged by Sydney Vines, then Professor of Botany, to help organise the department's own herbarium, a position that was officially recognised in 1895 when he was given the title Curator of the Fielding Herbarium. According to Arthur Church, when Keeble succeeded Vines in 1919 he encouraged Druce 'to teach what he knows', though Church witheringly added, 'if he knows anything worth teaching'.[25] In addition to his practical work, Druce wrote extensively about the Oxford herbarium and, as a 'thank you', was recommended in 1920 for a Fellowship of the Royal Society. The Fellowship was, however, awarded only in 1927, near the end of his life, and, despite his having earned an Oxford DSc by examination, he was never awarded the honorary doctorate he craved. The fact that he was made an honorary Doctor of Laws by the University of St Andrews, far away in Scotland, only heightened his disappointment.[26]

Druce's contribution to the Oxford herbarium was, unfortunately, marred by his inattention to detail for, characteristically, he carried out every task with too much haste. But the problems of the *living* Druce barely affected Tansley because, by the time the latter arrived in Oxford in 1927, Druce was 77 years old, his energies greatly diminished. Relations were cordial, at least on the surface, for Tansley wrote offering his warmest congratulations on the occasion of Druce's 80th birthday in May 1930.[27] It is unlikely that such warm feelings reached far below the surface for the two men were too different.

By the time of his death in February 1932, Druce had accumulated an estate worth £91 545 (equivalent to more than £4.5 m today), thanks to the sale of his pharmacy, wise investment, and astute money lending (most of the Oxford Colleges owed him money).[28] His will proved something of a bombshell, providing Tansley, among many others, with a host of difficult decisions. Once numerous small legacies were paid off, the remaining beneficiaries were to be the University of Oxford and the SPNR – the BEC was, surprisingly, almost completely omitted. Druce left his large house, Yardley Lodge, in Crick Road, north Oxford, together with his personal herbarium of over 200 000 specimens, and his library to the University on condition they should be kept together in order to serve as an institute of systematic botany, 'The Claridge Druce Herbarium and Institute'. It would focus on the British flora, rather than the tropical flora which was currently exciting taxonomists. His will also stipulated that his chosen curator, John Chapple, should be kept on, paid from Druce's estate.[29]

As the University's senior botanist, Tansley's opinions were central to the University's response to Druce's bequests. The Botany Department could expect to have a strong claim upon whatever money came to the University, thereby achieving a solution to its accommodation problems. The central problem for Tansley was that an institute of systematic botany, particularly one distant from the hub of botanical activity in Oxford, was not something he welcomed. He wanted to promote physiology and genetics, and through them ecology, not systematic botany.

The University wanted the cash, and was happy to accept property, but it too did not want to create an institute. In the event nothing happened quickly because Druce's will was contested when it become apparent that there had been not just one will, but many. Druce had kept changing his mind. Probate was finally granted only in 1936 when after lengthy legal action the University grudgingly accepted the package – to the disappointment of the BEC which had hoped to gain by default.; £19 000 was available for taxonomic research but could not be spent on new buildings.[30] Tansley was on the verge of retirement and about to be relieved of any responsibility for sorting out Yardley Lodge, which was left in limbo, seldom visited by him and barely used. Only its function as the continuing headquarters of the BEC remained. The house was finally given up by the University shortly after WWII. Druce's herbarium was eventually joined with the existing Fielding Herbarium when they were both moved to South Parks Road after new buildings for botany were opened there in 1952.

***

The bachelor-like life of a Don, 'living in' at an Oxbridge college, was a new experience for Tansley, who during his Cambridge years had gone home each night to Grantchester. Now his physical needs were met by college servants rather than his family and domestic staff. It was an experience he thoroughly enjoyed for, generally, he received a kindly welcome at Magdalen and 'was thus able to learn much more about College and University ways than would have been possible if he had taken a house in Oxford.'[31] Not all those ways were admirable. High Table and the Senior Common Room may have brought together the greatest intellects from across the spectrum of disciplines but, in the cloistered environment of a college, academic jealousies could often spill over into bitchiness and back-biting. In his diary for 31 January 1927, the English tutor and author of the *Narnia* books, C. S. Lewis, wrote, 'At dinner J.A. [philosopher John Alexander Smith, also a Magdalen Fellow] remarked that our new fellow Tansley (one of these professors who comes to us *ex officio* whether we will or no) "fancied himself as a psychologist" so we may expect good Socratic irony from J.A.'[32]

Tansley involved himself in the more positive aspects of Oxford's intellectual life, one glimpse of which is provided by the memoirs of Solly Zuckerman. The young South African had become friendly with Tansley's daughter Ka(therine) while completing his medical studies at University College Medical Hospital, where he was taught, among others, by Ka's future husband Dick (R. J.) Lythgoe. Invited by Ka to spend a weekend with her family at Grantchester, Zuckerman and Tansley quickly became friends despite a 30-year difference between their ages.[33] Tansley took Zuckerman under his wing. First he advised him to leave medicine in favour of an available post as research anatomist at the Zoological Society (London Zoo) where he could pursue his interest in primate evolution. Second, both before and after Zuckerman had moved on to the Department of Anatomy in Oxford, Tansley introduced him to some of the most brilliant brains in the University. Thus, at one dinner party Tansley held for Zuckerman at Magdalen, he was introduced not only to 'Harry' Weldon, Magdalen's distinguished senior philosophy tutor, but to two future Nobel Prize winners, Erwin Schrödinger, the theoretical physicist/philosopher, and Peter Medawar, a future pioneer of immunology and transplant surgery, but then still an undergraduate.[34]

After his move to Oxford, Zuckerman was able to effect some interesting introductions of his own.

Over a period Tansley and I dined *à deux*, alternately in Magdalen and at Museum Road [Zuckerman's house], in order to chat and taste wines … One day when it was my turn to have him [Tansley] for dinner, … I came home with Thornton Wilder [American playwright and novelist] … Even though I knew that Tansley often dried up in the presence of strangers, I asked Thornton to dine. Unfortunately, he arrived first, and when Tansley came he was taken aback to find that we were not going to be alone. Thornton at first did all the talking, having just returned from Vienna and Switzerland, where he had visited both Freud and Jung. He was enthusiastic about psychoanalysis and … started to describe the layout of Freud's study, laying particular emphasis on a picture of 'The David' which hung near Freud's desk. 'Do you see how the room was arranged?' he repeatedly asked Tansley. 'Do you see the picture?' 'Yes, I do', came the reply in the end, 'I gave it to him'. Up to that moment Tansley had hardly spoken, and certainly had given no hint that he had once been a disciple of Freud. After that the talk became easier.[35]

The Magdalen Philosophy Club provided Tansley with the welcome relief of an intellectual challenge far removed from the day-to-day problems involved in modernising Oxford's Botany Department. The interface between biology and psychology had long interested him but now a third element, philosophy, provided additional interfaces and further complexity. It also provided an academic battleground. On one side were the romantic 'idealists', led by J. A. Smith, who argued that truth about the real world could be attained only through studies in the history of thinking, and his ally, Robin George Collingwood, who believed a thorough knowledge of the history of ideas was a precondition for understanding the nature of scientific truth, which would ultimately lead to a revelation of the ultimate truth that is guaranteed by God's omnipotence and goodness. Collingwood was alarmed by the way that, as he saw it, Christian values were being sacrificed in favour of the sort of positivism and material realism he detected in Tansley's work.[36] Their opponents, the 'realists', included, in addition to Tansley, Charles Sherrington, who won the 1932 Nobel Prize for his research on nervous systems, and the young zoologist, John ('J.Z.') Young, who argued that in his discovery of nerve fibres in the squid and octopus he relied on physical facts about a real world and not some thing or concept incapable of being known in practice. The realists shared a mutual interest in the physical basis of psychology, and the latter's relation to philosophy. Archives of the Philosophy Club surviving in the University and Magdalen College libraries leave the collective impression that Tansley was not only an enthusiastic contributor but also, characteristically, the Club's de facto secretary.

Tansley certainly played his full part in Magdalen's life, its records showing that he rarely missed a college meeting, and that he served on a number of its committees. When The Grove, an area of college grounds once famous for its collection of (now ageing) elms, was attacked by Dutch elm disease in the late

1920s and was subsequently devastated by a severe storm on the night of 12 January 1930, it was to Tansley that the college turned for guidance. It accepted his recommendations, made in a report to the college meeting of 7 November 1934, in spite of them including radical plans which would involve significant alterations to avenues and vistas that had remained unchanged since the 17th century. What Tansley wanted to achieve was a 'park-like effect'. The Grove would be opened up by the felling of some elm and sycamore trees that had survived the storm and their replacement by fewer, and in most cases smaller, trees or shrubs, 'to create better views'. What may have helped win support was Tansley's stated desire, 'to see an increased use of the Grove by Fellows of the College'.[37]

***

One personal matter, stirring his deepest feelings, invaded whatever peace and academic seclusion Tansley found behind Magdalen College's ancient walls. There is little certainty about the identity of his 'beloved' or her movements, beyond the information that she went to South Africa during the 1910s. Whether she stayed there is not known but it is certain that in the summer of 1929 Tansley attended a meeting of the British Association for the Advancement of Science (BAAS) held in Cape Town and Johannesburg. The BAAS meeting clashed with an International Psychoanalytical Congress being held in Oxford and, since Tansley was the sole Oxford-based British member of the International Psychoanalytical Association, it might be expected that he would have been involved in organising or contributing to that meeting. The obvious inference if that there must have been something very important to take him to South Africa – his beloved. In one of Tansley's notebooks for this period, Cameron and Forrester have found the following undated entry:

> There is no 'armour' to protect one from such elemental hurts, I find. You must know this because the knowledge that we are sharing the pain may help. I am numb toward everything but these two days. But I cannot regret them, nor can I face the absolute finality … It eases the pain to write but it is an indulgence, and I have hurt you too much already, my very dear.[38]

Whether the beloved had remained in South Africa, or had gone there for the BAAS conference, their coming together seems to have been planned, for Tansley knew in advance that they would meet. Clearly, their feelings for each other were still strong and disturbing, but one or both had decided to break whatever ties still connected them, for they had no future together. The time had come to free themselves and move on.

***

Committed to invigorating his department and attracted to the intellectual life in and around his college, the flow of Tansley's publications slowed to a trickle during his Oxford years. Apart from a short booklet on Wicken Fen, describing work completed several years earlier, and a lengthy review of British beech-woods, written with A. S. Watt, his output was, with one exception, confined to book reviews and obituaries. That exception proved, however, to be the publication for which he is best remembered by his fellow ecologists, for in 'The use and abuse of vegetational concepts and terms' (1935) he launched the concept of the ecosystem. As so often in Tansley's life there was a South African connection – and it was critical.

Ecology had been quickly embraced by South African-born botanists, their ranks soon being swollen by immigrant botanists from Britain, such as Moss and Adamson. The South Africans argued that their country's diverse topography and climates, coupled with its uncontaminated habitats, offered the best research sites in the world. Some among their number, such as J. W. Bews, claimed a greater value for their research than that centred in heavily industrialised countries, such as Oliver's study of the effects of urban fog.[39]

In his capacity as Editor of the *Journal of Ecology*, Tansley received in 1934 a series of three lengthy papers written by the South African ecologist John Phillips, Moss's successor at Witwatersrand.[40] The papers dealt with 'Succession, development, the climax, and the complex organism', subjects close to Tansley's heart, but in his opinion presented as 'the pure milk of the Clementsian word'. In addition to Frederic Clements, with whom Phillips had spent a study period at the Alpine Laboratory,[41] the other influence on Phillips was the redoubtable Jan Christian Smuts, soldier, Prime Minister of South Africa from 1919 to 1924, expert on savannah grasslands, President of the BAAS in 1931, and originator of the term 'holism'. Like Clements, both Phillips and Smuts were known personally to Tansley. He knew Phillips from the Fifth International Botanical Congress held in Cambridge in 1930, where Phillips presented a paper, 'The biotic community' setting out an idealist foundation for ecological research. And he had met Smuts in 1929 when the latter was in Oxford, at the invitation of the Oxford idealists to receive an honorary degree and deliver three Rhodes Lectures on successive nights to a packed Sheldonian Theatre; frustratingly for the history of ecology, there is no record of their conversation.[42] For many years convinced of the wrongness of Clements' interpretation of the temporal changes affecting vegetation, Tansley could no longer hold his silence after receiving Phillips' three papers. While remaining polite to his would-be adversaries, his 'Use and abuse' paper showed Tansley at his most forthright.

What had upset him is easy to see. Phillips had claimed that his arguments were based on, and supported by, the ideas of Smuts, Clements, *and* Tansley! The last was certainly not the case.

Phillips, like Bews, was a disciple of the powerful Smuts, a man who took a keen interest in young South African botanists and who was in a position to advance their careers. After losing office in 1924, Smuts had devoted the next two years to writing *Holism and Evolution* in which he set out his philosophy. He proposed that everything in nature is interconnected. 'The new science of Ecology', he said, 'is simply a recognition of the fact that all organisms feel the force and moulding effect of their environment as a whole'.[43] The entire material world was alive, and organisms represented microcosms in which member cells cooperated in a regime of law and order, and which were in themselves models for human society. The life force in matter – a concept that appealed to the Oxford idealists – could best be observed in plants and animals. (Ominously, for he was opening the door to racial segregation in South Africa, Smuts proposed that some organisms are more 'significant wholes' than others, their creation being 'the meaning and trend of the universe'.[44]) While, in his Cambridge address of 1930, Phillips had performed a valuable function when he reminded ecologists that their studies of biotic communities should include both plants *and* animals – not either/or, as was so often the case – he continued, 'In accordance with the holistic concept of Smuts (1926), the biotic community is something more than the sum of its parts; it possesses special identity – it is indeed a mass-entity with a destiny peculiar to itself'. 'A biotic community behaves as a complex organism – in its origin, growth, development, common response, common reaction, and its reproduction'.[45]

If Phillips had used Smuts' holism to support Clements, in his papers submitted to the *Journal of Ecology* in 1934 he went on to argue, like Clements, that succession is always the result of biotic reactions, possibly accelerated or retarded by environmental factors. He added that succession is always progressive, and that its end-point is the 'climax community', or quasi-complex organism. There is ideally a single climax community for each regional climate.

The long friendship between Tansley and Clements was based on a deep mutual respect forged during their involvement in the IPEs of 1911 and 1913, but if Phillips thought this bond extended to their ecological beliefs he was disastrously mistaken.

The perfect platform for attacking Phillips, and Clements, was provided when Tansley was invited by the Ecological Society of America (ESA) to contribute to a *Festschrift*-both a conference in Ithaca, NY, and a commemorative volume of the ESA's journal, *Ecology*-celebrating the career of H. C. Cowles. Just as Tansley had from his studies of heath and fen in Britain rejected the idea of a 'monoclimax', in which all communities ended up as woodland, so too, long ago, had Cowles; in his case on the basis of the evidence of his PhD studies of dune systems in the area of the Great Lakes in the USA made between 1896 and 1898. (As he walked inland from the sandy shores of Lake Michigan,

Cowles typically passed through a series of communities in which the sand was progressively stabilised as the roots which colonized them died and created humus for a richer flora. However, he refused to believe that succession inevitably moved to a single climax for he had also observed retrogressive succession, confusing intermediary stages, and several kinds of climax. With the instability of dunes in his mind, he had written in 1901 that succession was 'a variable approaching a variable'.[46]) It was appropriate therefore that Tansley should use Cowles' *Festschrift* as his launchpad. In 'Use and abuse', he wrote:

> my conviction [is] that Dr Clements has given us a theory of vegetation which has formed an indispensable foundation for the most fruitful modern work. With some part of that theory, however, I have never agreed, and when it is pushed to its logical limits and perhaps beyond, as by Professor Phillips, the revolt becomes irrepressible …

and declared his intention was to be 'blunt and provocative'.[47] The (progressive) successions that lead from bare substrata to the highest types of vegetation in a climatic region, he asserted, are primarily autogenic; that is, the dominating factors of change depend directly on the plants themselves. Those (retrogressive) successions that lead away from the higher forms of vegetation are largely allogenic; that is, dominated by factors external to the plants. A climax is a relatively stable phase; it may be limited only by climate (climatic climax), or by soil type, grazing animals, fire, and the like.[48]

Totally refuting the Smuts–Phillips doctrine that 'holism' was a fundamental factor – a 'mysterious' factor, Tansley called it[49] – operative in the universe, Tansley conceded that mature, well-integrated plant communities (climaxes) had enough of the characters of organisms to be considered as *quasi-organisms*, but argued a much broader term was needed to describe an ecological unit. It should include the whole complex of plants and animals, which he called the biome, and all the effective factors of its physical environment. The term he chose was the ecosystem; it was 'a particular category among the physical systems that make up the universe. In [an] ecosystem the organisms and the inorganic a factors alike are *components* which are in relatively stable dynamic equilibrium'.[50] Ecosystems closest to equilibrium were the most likely to survive longest (modern ecologists may disagree[51]), although their inorganic factors or components made ecosystems generally less stable than chemical systems.

Citing the mathematician–philosopher Hyman Levy's recent book, *The Universe of Science*,[52] and drawing upon his own experience of psychology and interest in the philosophy of learning, Tansley acknowledged the ecosystem was in a sense artificial since it was imposed by man on nature. What it did, he said, was to reflect the fundamental need of the scientific mind to isolate units that it could recognise and comfortably deal with.

Use of the term was slow to take off but by 1988, the 75th anniversary of the founding of the British Ecological Society, members voted it first among the 50 most important concepts in ecology.[53]

Introduced to the world of ecology by Tansley who, in view of his detailed explanation and justification of it, has received the credit for it, the term 'ecosystem' was actually suggested to Tansley in the early 1930s by his young colleague, Clapham. This happened when Tansley asked him if he could think of a suitable word to denote the physical and biological components of an environment considered in relation to each other as a unit.[54] Tansley's failure to acknowledge publicly[55] his junior's contribution, albeit more acceptable behaviour in the authoritarian 1930s than today, is still disappointing. Clapham's family do not recall any resentment on their father's part, indeed his letters and all other evidence indicate he remained a close friend and admirer of Tansley. Clapham's attitude is explained not just by his generous nature, but also by the fact that the basis for the ecosystem concept, although not clearly formulated, was an old one, particularly among freshwater biologists. Thus, in his treatment of a small lake in the American mid-west as a microcosm, S. A. Forbes in 1887 came close in spirit to the ecosystem as he recognised the interactions involved in such closed systems. So too did the German limnologist, A. Thieneman in 1925, who adopted an old term, 'biocenosis', to describe the lake communities that he studied.[56] Clapham had, for the benefit of Tansley, refined and extended ideas that were already in the literature, but he had done so in a way that was immensely important for the future of ecology.

***

It would be easy but grossly unfair to think of Tansley at this stage of his life as a cold man, serious minded, austere, taciturn, and a manipulator of others. He could be each of those things at different times, and at any one time he did sharply focus his attention on whatever was his current interest, usually plant ecology. However, he could also be jolly, caring, and generous with both his time and his money. Whatever his mood, he invariably saw the good in those with whom he disagreed. Thus, over the years, he had had many disagreements with Clements about ecology, but this had not affected their friendship. When Clements died, Tansley wrote of him:

> The man who states a general theory which leads subsequent workers along the most fruitful lines of research performs a service which is fundamental to the progress of science, in that it helps to create the permanent structure of science, without which the amassing of detailed knowledge, and even the most brilliant

single discoveries, can have no coherent meaning. Such a theory may be overstated, it may contain flaws which make it unacceptable in its entirety; but if it also contains, as Clements's did, a general idea of the first importance on which subsequent advance can be based, its originator's name can never be forgotten.[57]

Tansley's caring side is illustrated by an incident that occurred shortly before he was due to pass the editorship of the *New Phytologist* to the younger editors. Clapham told his wife what happened:

> This afternoon he asked me to go to his room and gave me Godwin's review of James' book, which will appear in the first number of the New Phyt. [*sic*] after the change of Editorship (rather unfortunate, isn't it?). He has taken a lot of trouble over the review, because it was very harsh in its original form. Godwin altered it, and even then the Prof. had to soften certain 'asperities', as he called them. Today he wanted my opinion as to whether James would be hurt by it. He realises that James is sensitive …
>
> *A. R. Clapham to his wife, 19 January 1932*[58]

Two examples may serve to demonstrate his generosity. The first involves Lilian Clarke, who completed her degree in botany at UCL in 1893, taught by Oliver but just too late for Tansley. In 1896 she began teaching at the James Allen's Girls School, South London, where in her very first year she began to develop in its grounds 'The Botany Gardens'. Somehow her work came to Tansley's attention, possibly through the BAAS of which she was an active member, or possibly through the Linnean Society of which she became a Fellow. With his guidance and encouragement she changed what were simple systematic beds to ones whose plantings represented different ecological types: heath, bog, salt marsh, and sand dune. The plants were collected from their natural habitats and brought into school by the girls themselves. In her outdoor classroom Clarke taught from first principles, not from textbooks.[59] She was so successful that in 1917 she was awarded a DSc by the University of London for her thesis on botanical education. In the foreword to her book, *Botany as an Experimental Science in the Laboratory and Garden*,[60] published in 1935 after her death, Tansley paid tribute to 'her thoroughly sound fundamental ideas, her extremely clear and honest mind, her keen enthusiasm, and her indomitable energy and perseverance', regretting only that the book was as much a history as a sorely needed general textbook.

The second example offers a complete contrast for it involves a penurious Lincolnshire parson, the Reverend Edward Adrian Woodruffe-Peacock. An early member of the BES, Woodruffe-Peacock had already published 'A fox-covert study' in the *Journal of Ecology* when he and Tansley met on a Society field trip to Mildenhall in Suffolk in 1918. They became correspondents and, surprisingly since Tansley was an avowed atheist, friends. It emerged that

Woodruffe-Peacock had for many years been working on a detailed 'Rock–soil flora of Lincolnshire', in which he related the plants of the county to the various soils and environments in which they were found. The approach was so innovative and Tansley was so impressed by the quality of the manuscript that he generously offered to pay the £300 towards its publication costs. His offer was never taken up because Woodruffe-Peacock was either unwilling or too unwell to make the necessary revisions. Finally, Tansley wrote to Woodruffe-Peacock proposing that, since he had suffered a long illness, he should write up just a few species in detail, allowing Tansley to abbreviate and summarise the remainder. Woodruffe-Peacock's diaries show that he was unhappy with this plan, so only a minor part of the lengthy manuscript, that dealing with bracken, was ever published; the bulk of it passed unpublished to the archives of the library of the University of Cambridge.[61]

***

Tansley's achievements in Oxford clearly outweighed his failures for when he was due to retire in 1937 his immediate peers, the Biological Sciences Board, requested the University to extend his tenure for a further three years. The University refused but 'gave him £2900 from the University Chest, or its annuity equivalent (c. £300 a year) as pension, this being rather more than he expected'.[62] Unusually, the University allowed Tansley, the sitting tenant, to write the job specification for his successor. Naturally, it was a plant ecologist that the University was to seek next.[63]

Oxford's loss proved to be ecology and conservation's gain for in retirement Tansley completed *The British Islands and their Vegetation*, the great work he had begun in Oxford, served as President of the BES for a second time, and took a leading role in various war-time bodies that helped spawn the Nature Conservancy.

Notes

1. MacDonald, Prime Minister in 1924 and 1929–1931, was a science graduate who included botany in his studies at Birkbeck College, London. His daughter, Sheila, was a contemporary of Helen Tansley at Oxford; Sheila stayed with the Tansleys at Grantchester for lengthy periods in the early 1940s (Martin Tomlinson, personal communication). Chamberlain, Prime Minister in 1937–1940, was a member of the SPNR.
2. Sheail 1998, pp.6–7.
3. Oliver 1914, p.55.
4. Sheail 2002, p.124.

5. Ibid., p.126.

6. Renton D. 1999. Red letter days. Peak District, 24 April 1932. *Socialist Review*, issue 229.

7. Present on 20 January 1927, when Tansley was elected, were the Vice Chancellor (F. W. Pember), the President of Magdalen College (Sir Herbert Warren), Sir Charles Sherrington (Waynflete Professor of Physiology), Sir John Farmer (Imperial College, University of London), Professor Robert S. Troup (Imperial Forestry Institute, Oxford,) Professor Edwin S. Goodrich (Linacre Professor of Zoology), and Mr Applebey. (Oxford University Archives, Bodleian Library, DC 9/1/1, p.229). Among the unsuccessful candidates were Reginald Ruggles Gates (King's College, London), the first husband of Dr Marie Stopes, and Charles Edward Moss (University of Witwatersrand), Tansley's ex-colleague (Morrell 1997, p.236).

8. Porter 1968, p.45.

9. Gunther 1904, pp.5–7.

10. Tansley 1927, p.6.

11. Harrison1994, p.152.

12. Morrell 1997, p.236.

13. Ibid., p.239.

14. Ibid., p.233.

15. Tansley to the Vice Chancellor, 30 November 1930. Oxford University Archives, Bodleain Library, URB/BG/1 file 1.

16. Morrell 1997, p.240.

17. Family letters of A. R. Clapham (by kind permission of Jennifer Newton, née Clapham).

18. Ibid.

19. Ibid.

20. Ibid., 3 February 1932.

21. Tansley to the Vice Chancellor, 25 November 1935. Oxford University Archives, Bodleian Library, URB/BG/1 file1.

22. Morrell 1997, p.241.

23. Allen 1986, p.108.

24. Druce collected members' subscriptions into his own bank account, rather than setting up one for the BEC, paying from his own pocket the subscription of anyone whose name he felt would adorn the membership list, even though they had little interest in the BEC (Stephen Harris, personal communication).

25. Mabberley 2000, p.66.

26. Allen 1994, p.44.

27. Druce Memorial Volume, Archives of the Plant Life Sciences Department, University of Oxford.

28. Ibid.; Stephen Harris, personal communication.

29. Allen 1986, p.116.

30. Morrell 1997, p.243.

31. Godwin 1957, p.237.
32. Lewis 1991, p.442.
33. Zuckerman 1978, p.27.
34. According to Zuckerman, the dinner was held in 1932 but it was not until autumn 1933 that Schrödinger left Germany to take up a Fellowship at Magdalen. Possibly he may have been visiting Oxford in 1932.
35. Zuckerman 1978, p.92.
36. Anker 2002, p.612.
37. Brockliss 2008, p.700.
38. Cameron, Forrester 1999, p.89.
39. Anker 1999, p.36; Anker 2001, pp.68–9.
40. Phillips JFV. 1934. Succession, development, the climax, and the complex organism: an analysis of concepts. Part I. *Journal of Ecology* **22**, 554–571; Phillips JFV. 1935a. Succession, development, the climax, and the complex organism: an analysis of concepts. Part II. Development and the climax. *Journal of Ecology* **23**, 210–246; Phillips JFV. 1935b. Succession, development, the climax, and the complex organism: an analysis of concepts. Part III. The complex organism: conclusions. *Journal of Ecology* **23**, 488–508.
41. Phillips considered Clements 'the greatest living ecologist' and expressed surprise that Clements was not better appreciated in America. Clements, who found Phillips, 'the most promising of the younger ecologists', explained to his young admirer that local jealousies influenced contemporary judgements (Clements 1960, p.226).
42. Smuts frequently visited Oxford to stay with Margaret Clarke Gillet and her husband. Their friendship originated following the Boer War when, with her fellow Quaker, Emily Hobhouse, Margaret had helped relieve the sufferings of Boer women and children. Like Smuts, she was a keen and expert amateur botanist (Hutchinson 1946, p.vii).
43. Smuts 1926, p.349.
44. Ibid., p.109. Smuts believed there was a battle between two floras at the Cape. An ancient decaying flora was confronted by a younger, more vigorous, invading flora from the north. Black Africans were equivalent to the former; they had stopped evolving because of the African climate (Anker 2001, p.68).
45. Phillips 1931, p.20.
46. Cowles 1901, p.81.
47. Tansley 1935, p.285.
48. Ibid., p.306.
49. Ibid., p.298.
50. Ibid., 306.
51. Golley 1993, p.16.
52. Levy H. 1932. *The Universe of Science*. London: Watts.
53. Willis 1997, p.270.
54. Willis 1994, p. 81.

55. C. D. Pigott heard Tansley admit this while they were both staying at Clapham's home in Sheffield in 1952 (personal communication).

56. Lévêque 2003, pp.22–3; Willis 1997, p.268.

57. Tansley 1947, p.194.

58. Family letters of A. R. Clapham (by kind permission of Jennifer Newton, née Clapham).

59. Sanders 2005, pp.22–3.

60. Clarke LJ. 1935. *Botany as an Experimental Science in Laboratory and Garden.* Oxford: Oxford University Press.

61. Armstrong 2000, p.58; Seward 2001, p.62.

62. Letter of A. R. Clapham, dated 'August 1937'.

63. Morrell 1997, p.241.

10 The Magnum Opus, Grantchester, and Retirement

When by the mid-1930s Tansley felt that at last his Oxford department was running smoothly, his thoughts turned to writing a substantial book. 'He hesitated between a modern university textbook of botany … a history of the early development of Freudian psychology … and an expanded version of his earlier book *Types of British Vegetation* … he accepted the advice of his friends and made choice [*sic*] of the last-named project'.[1]

Those last years in Oxford were among the happiest in his life as, realising he needed to refresh his knowledge, he insisted on revisiting sites throughout Britain that were once familiar to him. He also visited new sites which others had told him held some special botanical interest. Sites close to Oxford could be visited during term time, often with colleagues. More distant sites were visited during vacations, each expedition requiring carefully planning, which invariably involved hostelries offering a warm bed, good food, and, of course, good wine. Tansley loved to be accompanied on these longer trips by family and friends but, in character, he led the party in the field, undaunted by the roughest terrain or the highest mountain in spite of being in his mid-sixties. His (usually) good spirits and energy are evident from the recollections of two of his most frequent companions.

> When he was collecting photographs [for *The British Islands and their Vegetation*] a party of us went with him to the Chiltern beech woods … On return he announced that it was his birthday and would we dine with him in his rooms in Magdalen saying that he had ordered all his favourite dishes. The first was hors

Shaping Ecology: The Life of Arthur Tansley, First Edition. Peter Ayres.
© 2012 by John Wiley & Sons, Ltd. Published 2012 by John Wiley & Sons, Ltd.

d'oevre banked up on its dish like a wedding cake in tiers – a speciality of Duke the
Magdalen chef. Then boiled bacon and broad beans followed by gooseberry fool
and stilton cheese. He produced also a jeroboam of champagne.

J. L. Harley, 1990[2]

In the summer of 1936, Tansley, accompanied by his wife, eldest daughter Ka,
son-in-law Dick Lythgoe (the party's photographer), and Roy Clapham, spent
August touring Ireland. The latter enthusiastically told his young wife about his
new lifestyle:

> The hotel at Tralee was very much de luxe, but quite reasonable in charges,
> I believe. We had salmon-trout for breakfast and grouse for savoury and the rest
> in that style! Tansley is paying for the whole party, by the way.
>
> *A. R. Clapham, 26 August 1936*[3]

Next summer, 1937, as Tansley was about to retire, he, Clapham, Alex Watt, and
others, toured Scotland, although this year their good spirits were occasionally
strained, as Clapham wrote to his wife:

> The biggest of these [forests] is Rothiemurchus, over the Spey from Aviemore,
> and we went there today. The Professor wanted to see the native pines near their
> upper altitude limit, and we found a hill (or mountain) where the map showed
> pines to 2000 ft [approx. 600 m]. There was a loch between the car's stopping
> point and this hill, and it was obvious that the shortest route was by the W. side of
> the loch, but the Professor insisted on going all round by the E. side because he
> went that way in 1910, so we were already tired, and it was already 3 o'clock,
> before we began our climb. In the end we reached about 1650 feet, missed our tea,
> and lost our usual sweetness of temper. I was most annoyed when I was leading
> on the descent and making straight for the W. end of the loch. Tansley suddenly
> turned off at right angles, got half way along the S. side, and then thought that my
> suggestion was better after all and went back to the W. end. I had to rush after him
> when he made his turn, and all to no purpose! The fact was he was very tired,
> having stumbled several times during the climb, and was also very thirsty through
> missing his tea. He brightened up after drinking water from the loch and smok-
> ing a pipe, and we got back to the hotel about a quarter of an hour late for dinner.
>
> *A. R. Clapham, 2 September 1937*[4]

Leaving Oxford in 1937, Tansley was able to settle quietly into domestic life
back in Grantchester. Soon the daily rhythms of the household were arranged
to fit around his working routines. During the day he would be in his study,
accompanied only by his ever present pipe. His grandchildren recall he never
lifted a finger around the house, save to plump the cushion of the chair in which
he had been sitting. A bicycle was sufficient for most of his purposes, a selection of extremely old-fashioned models being kept at Grove Cottage. When his

grandchildren stayed at the house they were kept well out of his way, daring only to intrude on his study to announce, with some trepidation, that lunch or dinner was served – an announcement hardly necessary because meals were always served on time. Food had to be piping hot. His habits became slightly eccentric. He avoided using running water for washing, and when he bathed he used a hip bath prepared by one of the servants. He avoided also the water closet in the house, preferring to trek to an earth closet at the end of the garden. His reasons, if they were ever given, have long been forgotten. Possibly, Lankester's teaching at University College, London (UCL) of the importance of nutrient cycling was ingrained in his psyche.

In the evening, after dinner, he often sat with Edith, playing cards but speaking little. Later they would listen to the news on the radio before he would get up, kiss Edith on the top her head, saying 'Goodnight, Fuss', and retire to bed. Many years had passed since he had, one family mealtime, so cruelly announced his infidelity. Now, if not entirely forgotten, the emotional pains were numbed by the passage of time, replaced by an elderly couple's need for comfortable companionship.

There was never any question that Tansley would stop working. Free from the distractions of leading a busy university department, he had the time and peace he needed to complete *The British Islands and their Vegetation*. This major commitment did not, however, stop him from agreeing to serve a second term as President of the British Ecological Society (BES) in 1938, though he did concede that he should give up the editorship of the Society's *Journal of Ecology*, after almost a quarter of a century. The presidency represented both an honour and an opportunity to pursue those of his ambitions for ecology that remained unfulfilled. Probably fortunately, he could have had at the outset no idea of the length of the path along which his re-commitment to the BES would take him.

***

Even before the stock of 1500 copies of *Types of British Vegetation* was exhausted, which was within three years of its publication in 1911, Tansley had confided to a correspondent, 'It should be revised rather extensively, but I don't want to do it now'.[5] A quarter of a century would pass before he found the time and inclination for the task. Over the intervening years, the old book had remained popular and, as Tansley was well aware, had 'fetched fantastic prices on the secondhand market'.[6]

Enormous advances had been made during the 28 years separating the two books, as Tansley summarised in his preface to *British Islands*:

> the study of British vegetation in the field, and to some extent in the laboratory, has made great progress. The knowledge of our natural and semi-natural plant communities is much wider and especially much deeper than it was in 1911.

In 1911 we wrote practically all we knew and a good deal that we guessed: and though many of our guesses were not far from the truth others have not unnaturally turned out to be wide of the mark. A new generation of workers has grown up, with deeper knowledge and better training than those of the small band of pioneers in the early years of the century … Without the results of their work … this book could not have covered the ground with any adequacy.

Those advances were embedded in academic journals not always easy for budding ecologists to find, and largely hidden from non-specialists. It was time to make the subject more accessible.

British Islands was Tansley's magnum opus. It has been described as 'a milestone',[7] literally and metaphorically. Weighing 2.2 kg, the single-volume first edition ran to 930 pages of text – later editions spared readers' aching arms by being split into two volumes. There were 179 figures, many drawn by Tansley's daughter Margaret, and 162 plates. Many of the 418 photographs had been gathered by the camera of Dick Lythgoe.

In Parts I (Environment) and II (History), Tansley set out what he called the 'background' of British vegetation; among his many aims, seeking to make the book more accessible to foreign readers and to emphasise the especially important part played in Britain by man's activities. Part III, 'The Nature and Classification of Vegetation', expounded his 'dynamic' view of vegetation, 'the explicit recognition that natural and semi-natural vegetation is constantly changing'.[8] The ecosystem was equivalent, he said, to a position of relative stability, or climax. Although he pointed out that British vegetation demonstrated time and again that, in addition to the Clementsian climatic climax, there were edaphic and biotic climaxes where succession had become more or less permanently arrested at what Clements would have called the 'sub-climax' stage (see Figure 8.1b). Thus, while oak forest might be the climax community of much of England, succession could be arrested by too little or too much drainage, leading to the development, respectively, of marshes in low-lying areas and mixtures of herbs and xerophyllous (drought tolerant) grasses on exposed ridges.[9]

Tansley argued, '"positions of equilibrium" are seldom if ever really "stable"', although 'recognition of "positions of stability" is a necessary first step in the understanding of vegetation. The more important sequel is study of the factors which maintain or disturb and often upset them'.[10] Ecosystems that developed under essentially similar conditions, and were dominated by the same life forms, belonged to the same 'formation type'. In the case of the deciduous summer forest type, the stabilising condition was a complex of climatic factors, while the pastured grassland type was stabilised by biotic factors. Reed swamp was a critical formation type because, while stabilised by water level, the normal accumulation of plant debris would lead to the accumulation of humus, a rise in soil level, and the invasion of land vegetation. Nevertheless, reed swamp was a legitimate formation type because it represented a position of relative stability, which might be maintained indefinitely under appropriate conditions.[11]

In passing, it is worth noting Tansley's emphasis in his later analysis of vegetation – if not in his analysis of the human mind – on *relative* stability. In doing so, he anticipated the views of the generation of ecologists that followed him, who, as they revealed the full complexity of ecosystems, found stability was rare. Instability was attributable to the constantly changing interactions between the physical and biological factors affecting ecosystems. Tansley the psychologist would have been particularly intrigued by the work of those ecologists, such as the brothers, Howard and Eugene Odum, who made the first attempts to quantify energy flow through ecosystems.

The remainder of Tansley's book was given over to the vegetation itself. By far the longest section was devoted to woodlands – the once dominant climatic climax of the country. Then came the grasslands, 'occupying by far the largest area of any British formation, for the most part an anthropogenic formation-type replacing forest as the result of grazing'.[12] Priseres (see Box 6.1) were represented by freshwater marshes, fens, and bogs, and, also, by maritime vegetation.

Although many of the locations described in detail would have been familiar to readers of *Types of British Vegetation* – the shingle beaches of Blakeney, the aquatic vegetation of the Norfolk Broads, the Pennine moors, and Scottish heaths – many were new. Since 1911, Tansley's travels had taken him to almost every part of Britain and Ireland. In collaboration with Adamson and Salisbury he had made pioneering studies of the chalklands of southern England, and he was able to draw upon the studies of his own students such as Jack Harley (on ashwoods), and of young colleagues such as Clapham (on the bogs, birchwoods, and pinewoods of Scotland). The work of old friends, like Godwin (on marshes and fens, such as Wicken) and Alex Watt (on beechwoods) was included, and all were gratefully acknowledged.

Tansley drew also upon the work of other young plant ecologists, with whom he had no strong personal connection. One such was William Pearsall[13] whose studies of the Cumbrian lakes, especially Esthwaite, formed the basis of two of Tansley's chapters. Pearsall was a native of Ulverston, a small town on the southern fringe of the Lake District, and it was said that he 'looked at lakes through the eyes of a plant ecologist, and saw them as stages in an evolutionary succession'.[14] He saw in his native lakes a series illustrating the evolution of glacial lake basins from a primitive type with a rocky floor, stony or peaty margins, and little or no inorganic silt, such as Wastwater or the higher tarns (small lakes), to the richly silted substratum which he found on parts of the shores of the more advanced lakes, such as Esthwaite.

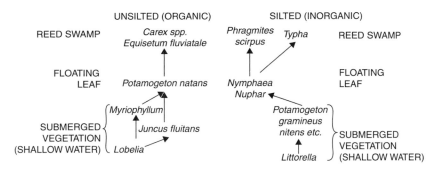

Figure 10.1 Contrasting hydroseres in shallow lake waters: based on Pearsall's researches in Esthwaite Water. (After Tansley 1939a.)

Begun before he volunteered to join the Royal Artillery in World War I (during which his hearing was permanently damaged), Pearsall's studies recommenced after the war – and after Tansley and Frederick Weiss had encouraged him to apply, successfully, for an assistant lectureship at Leeds University.[15] Pearsall's innovation was to start his studies with measurements of the physical characteristics of the water at different depths; this then enabled him to relate the vertical and horizontal distribution of plants to water chemistry and light penetration. (Light penetration into water decreases as depth increases and aquatic plants need 2% or more of full sunlight in order to grow. In the clear waters of Wastwater the critical 2% level was reached at ten metres depth, while in the silted waters of Esthwaite it was reached only at four metres).

Tansley told his readers how two distinct hydroseres had been described by Pearsall for the shallower waters of the lakes. Where silting was rapid and substrata inorganic, as at the northern end of Esthwaite, where Black Beck discharged its suspended matter, the 'silty type' succession occurred. It proceeded from *Littorella* (shoreweed) to *Potamogeton* (pondweed), *Nymphaea* (water lily), and then *Phragmites* (reed) and *Typha* (bulrush). In contrast, where silt was scarce, so that the shallow fringes of a lake were stony, and the substrata organic, the 'peaty type' succession was found, as in Wastwater . It proceeded from *Lobelia* (water lobelia) to *Myriophyllum* (water milfoil), *Potamogeton*, and *Equisetum* (horsetail) with *Carex* (sedge) (Figure 10.1).[16]

The main forces driving both hydroseres, or natural successions, were the increase of organic material in the substratum, and the raising of the soil level to the point where terrestrial plants could survive – the lakes and tarns were being 'filled in'. Whereas the silted series led to neutral or slightly alkaline 'fen' vegetation, the unsilted series led to acidic 'bog' vegetation. The latter thus paralleled the formation of peaty and acidic soils in moorland areas, a progression completed in the terrestrial 'raised mosses' common in the Lake District lowlands.

The most complete hydrosere known to Tansley had recently (in 1938) been observed by Clapham at Sweat Mere, near Ellesmere, in Shropshire. Using Clapham's unpublished data, Tansley described how, after domination by floating mats of *Typha*, *Carex paniculata* (greater tussock sedge) took over the Mere, forming one -metre high tussocks separated by standing water. As the *Typha* mat decomposed to black mud, *Salix atrocinerea* (common sallow) and *Alnus glutinosa* (alder) took over, the latter gradually forming a closed 'carr', or woodland. With the rise of *Betula pubescens* (brown birch) to co-dominance, the mud became firmer and grasses, such as *Agrostis stolonifera* (creeping bent) and *Holcus lanatus* (Yorkshire fog), replaced the last marsh plants. As drainage improved and soil pH continued to fall, *Quercus robur* (pedunculate oak), with some *Pteridium aquilinum* (bracken), came to dominate what was by now a recognisably dry land flora.[17]

Tansley went to considerable lengths to illustrate plagioseres, or deflected successions (Figure 8.1a). One example he used, Godwin's study of the effects of sedge cutting at Wicken Fen, was by 1939 well known to ecologists, but another example, the Broadbalk Wilderness at Rothamsted Experimental Station, was not. Broadbalk had long been interesting to agriculturalists because wheat crops had been grown there since 1843 and, stretching back to 1852, there was a unique set of records of the long-term effects on both crops and soils of applying fertilisers. From 1882 onwards part of the land had no longer been cultivated, the last wheat crop having been allowed to scatter its seeds naturally. The area became known as the 'Wilderness'; the untouched half developed into a dense thicket of oak–hazel woodland, with shrubs, while invading woody plants were periodically grubbed up from the other half. Lists of species growing in this latter area were assembled in 1886, 1895, 1903, and 1913, numbers of vascular plant species being, respectively, 40, 49, 57, and 65. In 1886, and also in 1895, dominance was shared by *Agrostis tenuis* (common bent-grass) and *Medicago lupulina* (black medick), both persistent weeds in the wheat crop, but in the survey of 1903 these had been replaced by *Dactylis glomerata* (cock's foot). By 1913 this had in its turn been replaced, for the survey conducted in that year by Winifred Brenchley and Helen Adam found that *Arrhenatherum elatius* (oat-grass) and *Centaurea nigra* (lesser knapweed) were dominant. Grubbing was nevertheless continuing to deflect the succession that would otherwise have led to oak (*Q. robur*) woodland (Figure 10.2).

When the Oxford colleagues Clapham and Baker visited the Broadbalk Wilderness 23 years later they commented on changes in the woodland area. 'Secondary woodland' was by then firmly established. Sycamores were now the largest trees, though some were dying, and the number of woody species had increased since 1913, when only oak, hazel, bramble, and ivy were recorded. There was an interesting comparison that Tansley was keen to point out between

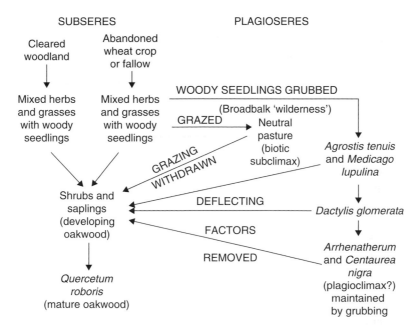

Figure 10.2 Scheme of some subseres and plagioseres, including that at Broadbalk Wilderness, leading to *Quercetum roboris*. (After Tansley 1939a.)

the rapidity with which young oak developed on the heavy loam (good soil) of Broadbalk and the much slower succession seen in the neighbouring, often waterlogged, Geescroft Wilderness, which involved an intermediate stage where the wet grassland was dominated by *Deschampsia caespitosa* (tufted hair grass).

***

Tansley's magnum opus has stood the test of time well. Inevitably, a few weaknesses have emerged in the seven decades since it was written. Clements' climax theory gave respectability to the tradition of stable, immensely ancient, 'primaeval forests' but doubts have steadily grown as to whether wildwood, isolated from human activities, has existed at any time in the 12 000 years since the last glaciation.[18] At the start of the Neolithic period, 3800 to 2000 BC, when the climate became drier and man increasingly spread and settled, the pattern of British woodlands was more complex than the endless, mixed oak forest, with alder and elm, Tansley envisaged. Birch and pine were more important in the north, and lime (*Tilia* spp.) more important in the south, than he judged, but the differences are a matter of degree, of refinement resulting from accumulating evidence.

Tansley's vision of more or less continuous, high, closed canopy forest has sometimes been challenged, most recently by Frans Vera,[19] who suggests high

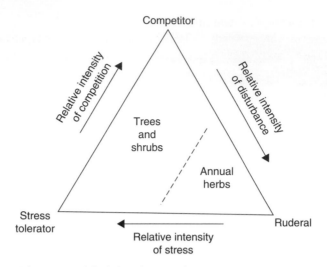

Figure 10.3   The *CSR* model of plant functional types. *S*(tress) equals phenomena that restrict photosynthesis, e.g. shortages of light, water, or minerals. *C*(ompetition) may be inter- or intraspecific. *R* stands for ruderals, plants favoured by disturbance, such as ploughing, mowing, herbivory, or fire. (After Grime 1977.[22])

forest alternated in both time and space with wood–pasture, a more open savanna-like system with significant areas of open grassland, maintained by deer and wild oxen. Tansley had little to say generally about the distributions of plants across landscapes where man was absent, and thought grasslands extended outwards, with man's help, from small areas of poor soil, such as recently deposited river terraces.[20] Vera's theory is in many ways analogous to Watt's gap-phase model for regeneration in high forest, proposed in 1947 in relation to British woodlands, although, on balance, Vera's evidence appears stronger for continental Europe than for the British islands.[21]

This is not the place to review modern ecological theory but, in passing, we may note the factors Tansley regarded as important in determining the nature of vegetation, namely, climate and soils, deflection (Tansley rarely used 'disturbance'), and competition, have been integrated by Philip Grime[22] into one widely accepted model. Vegetation at a real place and time is the result of an equilibrium established between competition, stress, and disturbance; thus, it can be represented at a characteristic position, or in a narrowly defined region, within a triangle, as is shown in Figure 10.3.

Such modern sophistication does nothing to diminish Tansley's achievement. It may have had its quirks – Tansley overlooked the importance of lime in the history of the vegetation of southern Britain, devoted more space to beechwoods than their extent merited, and underestimated the importance of man in their maintenance – but *The British Islands and their Vegetation* was a landmark publication. It brought together and distilled not only Tansley's own wealth of knowledge and experience but those of his contemporaries. Nothing so comprehensive had been attempted before, nor would it be attempted afterwards. The book said to plant ecologists, 'This is where we are', and to the wider world, 'This is what ecology is'. It did not shrink from pointing out uncertainties or areas that cried out for attention. It was written with the authority that can coalesce only around the unchallenged leader of a young science.

Notes

1. Godwin 1957, p.238.
2. J. L. Harley 1990, Archives of the New Phytologist Trust.
3. Family letters of A. R. Clapham (by kind permission of Jennifer Newton, née Clapham).
4. Ibid.
5. Allen 2010, pp.407–408.
6. Tansley 1939a, p.v; communist eastern Europe was culturally isolated from the west following World War II. In 1945, Tansley enlisted the help of the British Council when he wanted to send a copy of his *British Islands* to an eminent Polish botanist, Wladyslaw Szafer. They replied, 'we include it always in the standard literature we send abroad. ... we find it an invaluable work of reference and one of the first necessities in any serious collection of books on Britain' (Godwin 1977, p.17).
7. Allen 2010, p.407.
8. Tansley 1939a, p.vi.
9. Ibid., p.222.
10. Ibid., p.vi.
11. Ibid., p.vii.
12. Ibid., p.ix.
13. W. H. Pearsall was in 1938 appointed Professor of Botany at the University of Sheffield, and was responsible for establishing there a strong tradition of plant ecological study. He was succeeded by A. R. Clapham. J. H. Harley, one of Tansley's few research students in Oxford, also held a chair at Sheffield. Pearsall was later Quain Professor of Botany at UCL.
14. Sheail 1987, p.68.
15. Ibid.
16. Tansley 1939a, p.618.
17. Ibid., p.464.
18. Rackham 2006, p.83, p.105.
19. Vera FWM. 2000. *Grazing Ecology and Forest History*. Wallingford: CABI Publishing.

20. Tansley 1939a, p.164.
21. Rackham 2006, p.95.
22. Grime JP. 1977. Evidence for the existence of three primary strategies in plants and its relevance to ecological and evolutionary theory. *American Naturalist* **111**, 1169–1195.

11 The Years of Fulfilment, 1937–1953

The Nature Conservancy and nature reserves

Forgetfulness is often the companion of old age. When Tansley's notes went missing on the morning of the day he was due to deliver his Presidential Address at the Easter meeting of the British Ecological Society (BES) in 1939 his audience might have been excused if they had concluded their President's formidable mental powers were failing. Far from it, 'the meeting nevertheless received a very interesting extempore address'[1] on the subject 'The present position of British ecology', the published account showing Tansley was still perfectly up to date with current advances in ecology. As the next few years were to prove, the sharpness of his mind and his intellectual energy had never been stronger.

Near the end of his Presidential Address, Tansley touched on the subject of nature reserves:

> One of the most troublesome and irritating hindrances to ecological observations intended to serve as a basis for the study of successional change, and therefore having to spread over a series of years, is the liability to interference with or destruction of the vegetation of the area by events such as clearing, felling, draining, gravel digging, change of ownership, or 'development'. I myself have twice had serial observations brought to an end by events of this nature in areas which I thought were safe from interference.[2]

Shaping Ecology: The Life of Arthur Tansley, First Edition. Peter Ayres.
© 2012 by John Wiley & Sons, Ltd. Published 2012 by John Wiley & Sons, Ltd.

Their President also brought good news to the assembled ecologists:

> it was with great pleasure that the Council of this Society learned that the Forestry Commission proposed to reserve adequate areas which could be used for ecological research within some of [their] estates … and that they had invited the co-operation of the Society … I have long felt that similar opportunities were being missed in many of the properties of the National Trust.

Responding to public pressure for greater access to the countryside, which had resulted in increasing numbers of trespassers on its lands, the Forestry Commission was about to set aside for public use a number of Forest Parks, the first, in 1936, being the Forest of Ardgarten, by Loch Lomond.[3] Where the question of dedicated nature reserves was concerned, however, the Commission was reluctant to support them, fearing that they would inhibit its freedom to choose new forest sites.[4] As will be described in the first part of this chapter, the Commission was just one of a plethora of bodies – on the one hand including voluntary organisations and learned societies and on the other the ministries and departments of government (Table 11.1) – that aided and obstructed each other, replaced and subsumed each other, as the nation slowly and painfully edged towards the establishment of national parks and the Nature Conservancy.

Since the mid-1930s there had been a growing feeling that national parks were desirable, but possibly not affordable, that they should be administered by local authorities, and that reserves where wildlife could be preserved should be set aside *within* some parks. By the end of the 1940s, a national network of parks and *separate* reserves had been agreed, the parks, but not the reserves, to be administered by their own authorities, albeit in collaboration with local government. Most importantly for Tansley and the future of ecology, it was accepted that the reserves should be centres for research-based conservation.

* * *

It was a new body, The Council for the Preservation of Rural England (CPRE), founded in 1926,[5] that gave impetus to the whole process of change. The Council's aim was to coordinate the activities of various voluntary bodies and learned societies. It had 22 institutional members and many more affiliated bodies. The Council was akin to a pressure group; it would maintain an overview of planning applications that threatened the countryside, while encouraging landowners to allow the greater use of their lands for recreation.[6] By September 1929 the CPRE, together with its sister organisations in Scotland and Wales, had persuaded Ramsay MacDonald's government to set up a National Parks Committee (NPC) to investigate the desirability and feasibility of establishing one or more national parks for the protection of flora and fauna

Table 11.1 The path towards the establishment of national parks and the Nature Conservancy showing the contributions of voluntary bodies and learned societies, and of government.

Year	Voluntary bodies and learned societies	Government*
1926	Council for the Preservation of Rural England (CPRE) is formed to coordinate activities of voluntary bodies and learned societies	
1929–1931		Memorandum from CPRE prompts Prime Minister Ramsay MacDonald to set up the National Parks (or Addison) Committee: it recommends establishment of national parks
1934	Standing Committee on National Parks (SCNP) is set up by the CPRE	
1940		Winston Churchill asks Lord Reith (soon succeeded by Lord Jowitt) to plan for post-war reconstruction. Scott Committee starts work
1941	Society for the Promotion of Nature Reserves (SPNR) convenes the Conference on Nature Preservation in Post-war Reconstruction (CNPPR)	
1942	The CNPPR forms the Nature Reserves Investigation Committee (NRIC). The British Ecological Society's Nature Reserves Committee (NRC) also set up (chaired by Tansley)	Scott Committee recommends nature reserves within national parks. Jowitt asks Dower to make on the spot assessments of potential sites for national parks. Douglas Ramsay later asked to perform a similar task for Scotland
1943	NRC's report, 'Nature conservation and nature reserves', published. Tansley writes 'Nature reserves' for *The Spectator*	
1945	Tansley publishes *Our Heritage of Wild Nature*	Dower and Ramsay report. Hobhouse Committee set up, with associated Wild Life Conservation Special Committee (chaired by Huxley, then Tansley)

(Continued)

Table 11.1   *(cont'd)*

Year	Voluntary bodies and learned societies	Government*
1946		Ritchie Committee is set up for Scotland, to parallel the work of the Huxley Committee
1947		Hobhouse and Huxley Committees report
1949		Nature Conservancy (chaired by Tansley) is created by Royal Charter. It has statutory powers under the National Parks and Access to the Countryside Act

*1940–1945, wartime coalition; May to July 1945, Conservative (caretaker); 1945–1951, Labour. More detail is provided by John Sheail's *Nature in Trust* (1976).

and improvement of recreational facilities for the public. Under the chairmanship of a junior minister, Christopher Addison (Table 11.1), the committee took two years to receive evidence and prepare its report,[7] which favoured the establishment of a series of parks, to be administered by statutory authorities for England and Wales, and for Scotland. Unfortunately, publication of such bold proposals coincided with the worst of the economic depression, described in Chapter 9; the proposals had to shelved, despite MacDonald's personal sympathy, because his government was understandably forced to give them low priority. Following their huge disappointment, the voluntary bodies regrouped, among other things forming in 1934 a Standing Committee on National Parks (SCNP) which would maintain pressure on government and generally publicise the case for national parks.

While, in the broadest terms, there was among the bodies campaigning for national parks a shared ambition for the increased recreational opportunities they would provide, there was also a recognition that achievement of their common aim would in itself create problems to which there were as yet no agreed answers. Thus, if their potential was to be fully exploited, national parks would require better public access, which might mean building new roads and, even, railways. Also, while increased employment opportunities for people living within the parks was generally thought to be a good thing, the growth of businesses, new and old, would have to be carefully controlled, and planning regulations strengthened. This would unavoidably impinge on the freedoms that farmers and major landowners had previously enjoyed – which was, probably, a bad thing. And, most serious of all for those interested in protecting flora and fauna, let alone in studying it scientifically, increased public access to the land would inevitably threaten the very existence of those fragile communities and habitats that the parks should be protecting.

Experiences from countries where national parks had already been established were not helpful. The world's first recognisable National Park had been opened in 1872, at Yellowstone in the USA,[8] but, as in South Africa where the Kruger National Park was opened in 1926, the protected areas were vast, and human population densities very low. The first parks in Europe were set aside in 1909 when Sweden designated nine areas representing distinct landscape types, the majority in remote areas well away from major towns and cities. Most European countries, including Britain, had until the 1940s preferred to use legislation to protect their wildlife, and there was still considerable support for this strategy in Britain. If national parks were to be established successfully in Britain – a country with a relatively small land area and high population density – a host of conflicting interests would first have to be resolved.

Many men and women helped facilitate progress towards the establishment of national parks and the Nature Conservancy, but, where Tansley's involvement was concerned, none was more important than Max Nicholson or Julian Huxley.

Nicholson was 'a difficult man … difficult to keep up with. His colleagues found him exhausting. He was everywhere. His whirlwind of ideas and demands taxed all around him'.[9] An Oxford-educated historian, who in his student days had made himself an expert on bird life, he became a leading light in the establishment of the British Trust for Ornithology. For pleasure he wrote about birds, but to earn a living he took the post of Deputy Editor of the *Week-End Review*, writing mainly about politics. At the start of the 1930s there was widespread pessimism linked to a feeling that unfettered capitalism had led to the chaos and depression that was affecting the developed world. A feeling was rife that progress could be made only through planning. Addressing the Labour Party's annual conference in 1930, Ramsay Macdonald proposed the cure for the breakdown of capitalism was organisation, adding 'organisation that will see to it that when science discovers and inventors invent, the class that will be crushed down by reason of knowledge shall not be the working class, but the loafing class'.[10]

Moderate in its views was an organisation known as Political and Economic Planning (PEP), formed in response to an article, 'A national plan for Great Britain', written by Nicholson and published in February 1931 as a supplement to the *Week-End Review*. Supported by an annual grant of £1000 from Leonard and Dorothy Elmhirst[11] (founders of Dartington Hall in Devon, a centre for progressive education), the PEP called for the reorganisation of the country's economic, social, and political organisations in ways consistent with its liberal traditions and individual freedom. PEP's influence was out of proportion to its size. At its beginning in 1931 it had 50 members and by 1939 this number had grown only to approximately 150. What PEP did very successfully was to bring together, if only transiently, many of the progressive thinkers of the day. It was through the PEP that Nicholson was able to meet influential figures such as Lord Reith (ex-Director General of the BBC), and leaders of the amenity movement, such as John Dower and Clough Williams-Ellis, each of whom would contribute to the progress towards the establishment of national parks.

Nicholson's talents were recognised when he was recruited for the wartime civil service. Following the end of the war and the election of a Labour government, Herbert Morrison, the Deputy Prime Minister picked out Nicholson to head the Office of the Lord President of the Council. As Chairman of the Lord President's (Cabinet) Committee, which dealt with the government's domestic legislative programme, Morrison, and therefore Nicholson, exerted considerable influence over Cabinet decisions and the content of individual pieces of legislation. Morrison's portfolio of responsibilities included not only scientific research – which he believed could play a pivotal role in Britain's recovery from war[12] – but, more importantly in the context of national parks and nature reserves, he was the conduit for the Minister for Town and Country Planning, Lewis Silkin. Silkin, who was not a Cabinet Member, depended on Morrison for parliamentary time. Moreover, Morrison, or in practice his senior civil servant, Nicholson, was procedurally involved in the drafting of any legislation coming from Silkin, such as the landmark National Parks and Access to the Countryside Bill of 1949.

At the core of the PEP was a small directorate that included Julian Huxley, though only briefly because his political sympathies were more to the left. For a short period in the 1930s, Huxley and Bertrand Russell produced their own monthly publication, *Plan*, calling for 'a scientifically planned economic environment, but also the planning of education, law, landscape and population'.[13] Huxley's wife once said to him, 'So many fingers in so many pies. What a pity you haven't got more fingers'.[14] Julian was the grandson of Thomas Henry Huxley and the brother of Aldous, author of *Brave New World*. After Eton School and a first class degree in zoology from Oxford, Julian, sometimes headstrong and always restless, moved from job to job and country to country. Interested in animal behaviour, especially that of birds, he was by 1925 occupying the Chair of Zoology at King's College, London, but only two years later gave this up to concentrate on *The Science of Life*, a massive book he was writing in collaboration with H. G. Wells. The success of the book persuaded him to live as a popular writer, broadcaster, and lecturer, only to give this up in 1935 when he took up the post of Secretary of the London Zoological Society (where one of his research collaborations included Solly Zuckerman and Bertrand Russell).[15]

The harsh reality was that, despite the vogue for planning, and despite reports such as that of the Addison Committee and the work of committees such as the SCNP, realisation of the dream of national parks was as far away at the end of the 1930s as it had been at its beginning. It was World War II that would prove to be the catalyst for change.

As German U-boats blockaded British shipping lanes, the country was forced to consider ways of becoming more self-sufficient in food and timber production. There quickly developed a direct threat to the countryside and to nature conservation. Meadows were ploughed, marshes were drained, and peat was extracted from bogs. More and more land was taken over for military training, and there was no sparing even fragile communities, such as those of the sand dunes at Braunton Burrows, north Devon (used for rehearsals of the Normandy 'D-day' landings), or the chalk grasslands of Kingley Vale (used for artillery training). But these events happened piecemeal and their impact was gradual. It was the dramatic blitz of London and other major cities, in September 1940, that prompted Winston Churchill to ask Lord Reith to draw up a plan for reconstruction that would not only address the short-term repair of the cities but would look ahead to post-war reconstruction in general (see Table 11.1). Within a month the industrious Reith had drafted a Cabinet Paper advocating a national plan that included a limit to urban growth, preservation of fertile areas of farmland, and the protection of areas of natural and historical interest.[16]

Apart from being promptly redirected by Churchill to concentrate more on the immediate problems created by war, Reith had, as an outsider to government, naively failed to recognise the strongly territorial behaviour of the ministries responsible separately for industry, health, agriculture, etc. His proposals produced no action, therefore, except that he appointed another committee, this time A Committee on Land Utilisation in Rural Areas, under the chairmanship of Lord Justice Scott, to look into the impact on the countryside of new industry, and the associated housing it would require. The Scott Committee's report was not completed until 1942. Meanwhile, the voluntary bodies and learned societies were mobilising their expertise and resources. Tansley was about to become involved.

In the summer of 1941, and thanks to funds newly acquired through the Druce bequest, the Society for the Promotion of Nature Reserves (SPNR) convened a 'Conference on Nature Preservation in Post-war Reconstruction' (CNPPR) (see Table 11.1).[17] Its outcome was a memorandum for submission to government that was to prove both controversial and a watershed. The drafting committee, which included E. J. Salisbury (since 1929, Quain Professor of Botany at University College, London), agreed that while much attention had been given to the creation of national parks, too little had been given to the creation of nature reserves. They proposed that reserves must be part of any national planning scheme, and that there should be an official body, representing scientific interests, to draw up detailed proposals. They added that the running of such reserves should be vested in those individuals with practical expertise in wildlife management. It was the issue of control that was once again the source of disagreement, with the ambition of the County Councils' Association to control any national parks almost forcing the SCNP to withdraw from the Conference. A compromise was reached and, thanks to the intervention of Julian Huxley among others, the SCNP stayed in the Conference. However, 'any sense of rapport between the national-park and the local-authority interests within the "Conference" had been lost. The clash led to an even greater concentration on "the scientific aspects of nature preservation"'. From this point onwards there was discernible among conservationists, 'A yearning to break free from the amenity and recreational aspects'.[18]

The Conference's memorandum was published in November 1941 and in January 1942 its representatives were included in a deputation that met with Lord Reith to discuss national parks. However progress was soon hampered when Reith lost office, to be replaced by Sir William Jowitt and fresh thinking. Like Reith, he accepted the need to include nature reserves in post-war planning but he pointed out that, if they were not to be given low priority, the voluntary bodies and learned societies must provide government with hard evidence that would persuade it otherwise. The Conference accepted the challenge and in June 1942 set up the Nature Reserves Investigation Committee (NRIC) to carry out the task.[19] Two months later the long-awaited Scott Report was finally published. Within it was the suggestion that 'some of the National Parks will naturally form or contain nature reserves'. Scott's report was already out of date for the debate had moved on. Nature reserves had become a *separate* issue. Reflecting the altered mood of the times, and even before Scott reported, Jowitt had commissioned John Dower to go out from London and examine *on the spot* the practicalities of setting up national parks in places such as the Lake District, the Peak District, Snowdonia, and Dartmoor.

Except for its chairman and a representative of the Forestry Commission, all members of the NRIC were scientists, among their number being E. J. Salisbury, W. H. Pearsall, and John Ramsbottom. The first two were also members of the Nature Reserves Committee (NRC), set up by the BES and chaired by Tansley. The NRC had been asked to identify potential national habitat reserves for their use as outdoor laboratories and classrooms (Table 11.1), a major biological initiative for which Salisbury, in particular, had been lobbying hard within the Royal Society. Also sitting on both committees, and further strengthening the links between them, was Cyril Diver. His experience as Clerk to the Committees of the House of Commons had led to him being invited to be Drafting Secretary for the Conference; thus, Diver was a man who connected several bodies (he was, furthermore, President of the BES between 1940 and 1942, having distinguished himself by his researches on the ecology of the South Haven Peninsula in Dorset).

Tansley was the BES's obvious choice of Chairman of the NRC for he was *the* senior figure in British ecology and author of the definitive *British Islands and their Vegetation* – for which he had been awarded the gold medal of the Linnean Society. Equally important, ever since the formation of the British Vegetation Committee in 1904, he had won a high reputation as a wise and skilful chairman (Chapter 7). His first success was to win the goodwill of Herbert Smith, Secretary of the SPNR and its NRIC, who was initially doubtful about the value of the BES's contribution. A survey of the BES's 360 members was then organised, asking for information and opinions regarding the plants, animals, and habitats that should be protected, and the methods by which this could be achieved. After balancing such opinions, it was proposed there should be 49 national habitat reserves – some, such as at Braunton Burrows and Kingley Vale, being described as of 'outstanding importance' – and 33 scheduled areas where development inimical to wildlife should be banned or severely restricted. On the list were a further eight sites, such as Wicken Fen and Blakeney Point, where wildlife was already protected. The list was submitted to the NRIC.

Guided by Tansley, it was agreed in April 1943 that the remit of the BES's NRC should be extended so that it should not only review the arguments for nature conservation and assemble the best case for support, but also explain how ecologists should be involved in the management of conservation areas. Conservation was now firmly established in the BES's thinking and planning.

One month earlier, in March 1943, the NRIC had produced its report, *Nature Conservation in Great Britain*. (Unsurprisingly, since it shared several members with the BES's NRC, its list of reserves and scheduled areas had much in common with the latter's list.) Tansley's committee was thus able to carry out its deliberations in the light of the NRIC report. Their memorandum, 'Nature conservation and nature reserves' was ready for publication in the *Journal of Ecology* in May 1944, and 750 offprints were either distributed to interested parties or sold to the public.[20] The BES's report broadly supported the NRIC's proposals, the main point of disagreement being that the BES suggested, 'the division of Reserves proposed by the N.R.I.C. into Habitat, Species, Amenity and Educational Reserves should not be sharply defined and rigorously carried out'.[21] The BES's document presented a detailed argument in favour of a much simpler scheme that recognised only two categories of nature reserve, National and Local. The BES's report ended, 'your Committee recommend the establishment of a National Wild Life Service ... whose functions would include continuous research as well the administration of National Nature Reserves ... an independent [body] under the Privy Council', a body comparable with the

Medical Research Council. Here was another vital difference between the NRIC and the BES. The former proposed vesting control in the hands of a 'National Reserves Authority' that would be a department of the Ministry of Town and Country Planning. The latter feared such an arrangement would compromise *independent* research.

When the Dower Report was published in 1945, it shared the NRIC view. Predictably, Tansley disagreed because the Dower Report's position was at odds with his fundamental beliefs. The independence of the research worker was something he had always believed in passionately and, as will be seen in the final chapter, was something for which he worked actively, among other things helping to found in 1940 the Society for Freedom in Science. Deciding to fight for his beliefs and to keep up the pressure on government, he wrote an article, 'Nature reserves', for the *Spectator* magazine (1943),[22] and he published in 1945 a small, beautifully illustrated book, *Our Heritage of Wild Nature. A Plea for Organised Nature Conservation.*

G. M. Trevelyan had in 1929 used his status as the country's best known historian to launch an emotional appeal for the preservation of the English countryside; an appeal in which he castigated politicians for their inactivity.

> To statesmen thinking of the near future of their country, the preservation of natural beauty ought to be an anxious care. Yet, though the leading statesmen of all parties support as individuals, as politicians they have, hither to, done nothing, nor, I suppose, will they until public opinion stirs more actively in this manner.
>
> G. M. Trevelyan, *Must England's Beauty Perish?*[23]

Openly reminding his readers of Trevelyan's question, and arguing that the threat to the countryside had become even more imminent and dangerous since 1929, Tansley proposed in *Our Heritage* that 'planning for the preservation of rural beauty must be directed to the deliberate conservation of much of our native vegetation, since this is an essential element of natural beauty'.[24] Beauty, he said, depended largely upon vegetation; ecology lay at the foundation of those industries (agriculture and forestry) that depend on the management and use of vegetation; and education should include a knowledge of the nature and significance of different kinds of vegetation (and the animals that live in them) for this would 'widen the new citizen's intelligent outlook on the world, would quicken his feeling for the Britain he has inherited'.[25] After describing some of the most spectacular examples of England's natural beauty, he put the case for a National Wildlife Service. *Our Heritage* was the first of a number of publications Tansley produced late in his life in which he was at pains to simplify and popularise ecology. 'What is wanted', he wrote, 'is unequivocal public recognition of nature conservation as a national interest'.[26]

Despite more political infighting, and his own misgivings, the Coalition's Minister for Town and Country Planning, W. S. Morrison, responded to the Dower Report and the wishes of the Cabinet's Reconstruction Committee by establishing the National Parks Committee. The NPC became known as the 'Hobhouse Committee', after its chairman, Sir Arthur Hobhouse, the Chairman of Somerset County Council. The composition of the Committee is charmingly described in the relaxed memoirs of one of its members, Julian Huxley.

> Later in that summer of 1945, I was put on the 'Hobhouse Committee on the National Parks for England & Wales'.
>
> Max Nicholson, permanent secretary in the Lord President's Office, was responsible for setting it up and had chosen an excellent team including Sir Arthur Hobhouse as chairman, a wise and witty county magistrate with much experience in local administration;
>
> Leonard Elmhirst, with his knowledge of Dartmoor, of his own planned estate, and of the problems of access to wild country; Pauline Dower,[27] the 'strong-minded daughter' of Sir Charles Trevelyan in Northumberland; Professor Tansley of Cambridge, the leading plant ecologist in Britain; myself, with general biological knowledge and experience of national parks in Africa; Clough Williams-Ellis, and one or two others, with Max as secretary.
>
> Hobhouse was responsible for the National Parks side while I was made chairman of the Ecological Committee, which dealt with the problems of nature reserves and general ecological damage – and its prevention.
>
> After formal meetings in London we set out on a round of visits to possible sites for parks and reserves. We went to the Yorkshire dales and Pennine moors. [They went also to the Scottish borders, the Brecon Beacons, and Pembrokeshire].[28]

What Huxley casually called 'the Ecological Committee' was officially named the 'Wild Life Conservation Special Committee', though it too became widely known after the name of its Chairman, in this case the 'Huxley Committee'.[29] Against opposition from the Ministry of Town and Country Planning, who tended to see wildlife conservation as an irrelevance, Dower had pressed hard for such 'a committee of first class scientists', and he had prevailed. Some of the scientists had served on the earlier BES Nature Reserves Committee, chaired by Tansley. Cyril Diver, who helped draft the report of the Huxley Committee, along with its Secretary, Richard Fitter, had been a member of the NRIC – where, similarly, he had a major role in drafting its reports – but most members of the NRIC were overlooked, to their annoyance.[30] The committee should probably have been called the Tansley Committee for, although Huxley signed the final report, the bulk of the chairmanship fell upon Tansley after Huxley's wanderlust took him in May 1946 to the United Nations in New York (where before long he was made Director General of the United Nations Educational, Scientific and Cultural Organisation).

Max Nicholson had been a major influence within what in later years he was pleased to call 'a Think Tank' with, in his view, 'Tansley the effective leader'. The Secretary, Richard Fitter, was, like Nicholson, a distinguished, self-taught ornithologist; the two had worked together for several years after Fitter joined the research staff of PEP. The team gelled perfectly; 'decades were to pass', Nicholson wrote, 'before its Report showed any sign of ceasing to serve as a positive and reliable guide to the way ahead'.[31] (Comparable progress in Scotland had been slow and it was only in January 1946 that the Ritchie Committee was set up. Nevertheless, under the chairmanship of James Ritchie, it worked swiftly and prepared an interim report that was submitted alongside that of the Huxley Committee.)

Nicholson had pulled all the right strings. He had ensured not only that the recommendations of the Hobhouse and Huxley Committees were compatible but that the two had proceeded at a similar pace. This meant that in 1947 their recommendations came in tandem to Herbert Morrison, thereby increasing the chances that recommendations of the Huxley Committee (some would say, the less nationally important of the two) were not lost. Morrison's support was readily gained. Steered by Nicholson, he submitted his own proposals to the Privy Council in April 1948 and, in February 1949, he announced to parliament that the Nature Conservation Board and the Biological Service, as recommended by the Huxley Committee, would be subsumed into a single body, the Nature Conservancy. Tansley would be its first Chairman and Diver its Director General. What Nicholson called Tansley's 'wise and patient work' had at last born fruit.[32]

The Nature Conservancy would have scientific management at its heart. Its statutory powers came from the National Parks and Access to the Countryside Act of 1949[33,34] and its royal charter defined its responsibilities thus:

> to provide scientific advice on the conservation and control of the natural flora and fauna of Great Britain; to establish, maintain and manage nature reserves in Great Britain, including the maintenance of physical features of scientific interest; and to organise and develop the research and scientific services related thereto.[35]

The Conservancy would in effect function as a research council – just like the Medical Research Council, as Tansley had hoped (p. 164). In 1965, it evolved into, and was renamed, the Natural Environment Research Council.[36]

Given the task of identifying the first parks (and Areas of Outstanding Natural Beauty (AONBs)), the work of the National Parks Commission was encumbered by the legal requirement imposed on it to work with local authorities and Silkin's Ministry of Town and Country Planning.[37] In contrast, and crucially, the Nature Conservancy did not work under such constraints as it identified National and Local Nature Reserves.

* * *

Richard Fitter's description of the Hobhouse Committee, like Huxley's, suggests a familiarity, a cosiness, among its members.

> Some influential behind-the-scenes work, in the typical British establishment fashion and no doubt largely over lunch in the Athenaeum,[38] by a group which included Julian Huxley, A. G. Tansley, Max Nicholson, Cyril Diver and Sir John Fryer, the Secretary of the Agricultural Research Council, resulted in the appointment of the Wild Life Conservation Special Committee, under Huxley's chairmanship, as an appendage to the Ministry of Town and Country Planning's Hobhouse Committee on National Parks … To this Wild Life Committee I was appointed the very junior secretary.[39]

As a highly esteemed scientist, whose name leant respectability to scientific ecology and wildlife conservation, Huxley was at the centre of a web of connections which permeated the committees. He had known Arthur Hobhouse for many years and was a close friend of both Leonard Elmhirst and Clough Williams-Ellis. Huxley was *related* to Pauline Dower – literal familiarity – for his mother was the aunt of Janet Trevelyan (née Ward), wife of G. M. Trevelyan, whose brother, Charles,[40] was Pauline's father.

Tansley had known Max Nicholson since the 1920s, and was well acquainted with the biologists on the Wild Life Committee. John Gilmour had recently collaborated with him in a lengthy, and ultimately unsuccessful, attempt to produce a New Students' Flora. (As Chapter 12 reveals, Tansley regarded highly Gilmour's expertise and diplomacy.) E. B. Ford and Charles Elton, who had both been taught by Huxley at Oxford, were Fellows of Oxford colleges when Tansley arrived at Magdalen. Whereas Ford was somewhat eccentric and difficult to work with,[41] Elton, who had a fervent hatred of committees, was admired by Tansley; a feeling that was fully reciprocated.[42] Elton had founded the Bureau of Animal Population in Oxford in 1932 and, in the same year, launched the *Journal of Animal Ecology* for the BES. In his last years in Oxford, Tansley became excited by collaborations he was developing with Elton, animal ecology's young leader.[43] It is noteworthy – because it casts light on the early struggles of animal ecology – that although the Huxley–Elton relationship had sometimes been strained, the two men had been comrades-in-arms for, as Huxley explains in his *Memories* where Elton's letter below is published, 'in the early 1920s ecology was a new subject, not fully approved of by "classical zoologists"'.[44]

It was mainly you and your teachings and later your book that kept my interest in evolution alive … I had meant to say to you publicly how much I owed to you in the difficult early days of my ecological work.
Charles Elton to Julian Huxley, December 1970.

Cyril Diver was pivotal to the success of the Wild Life Committee. Already known to Tansley through his work as drafting secretary for both the CNPPR and NRIC committees, and his Presidency of the BES, Diver was the committee member every chairman needs. In Diver's obituary, published in *The Times* on 28 February 1969, Huxley wrote, 'He distinguished himself by his passion for thoroughness … and his readiness to take away a tangle of facts and opinion and to bring back a cogently stated and usually acceptable draft'. An anecdote of Elton's nicely illustrates both Tansley's style of chairmanship and his reliance on Diver.

There was (some years later) a meeting of the Nature Conservancy at which an unnamed company, making a bid for a contract to spray roadsides with herbicide, referred to the English Rose as a noxious weed. 'Do you realise, gentlemen', I [Elton] said, 'that you are about to destroy the basis of much of our English poetry …?' Tansley sat up and said, 'H'm'. Then he turned to Diver and said, 'Can't you do something about this, Diver?' The problem was resolved through the Establishment network, [Elton adding] 'Many awkward situations have been thus resolved during luncheon at the Athenaeum'.[45]

The Hobhouse and Wild Life Committees may have been riddled with what today would be deemed unacceptable 'cronyism', yet within their congenial atmosphere Tansley and his colleagues worked effectively. Quite simply, they persuaded the government to establish a body that would conserve British wildlife.

* * *

When still a relatively young man, in 1913–1914, Tansley had been one of a small group of botanists whose suggestions for nature reserves were incorporated in the 'Rothschild list', drawn up for the SPNR. His three recommendations for England were Kingley Bottom (as the Vale was then known), Staffhurst Wood, Surrey, an example of a Wealden wood with mature trees of beech, hornbeam, yew, and oak, where coppiced areas had been allowed to regenerate naturally, and Wistman's Wood, an ancient oak woodland 400 metres high on Dartmoor having an exceptionally rich epiphytic flora of mosses, liverworts, and ferns.[46] An example of *Quercetum roboris* (see Figure 10.2), formed under extreme conditions, Wistman's dwarf pedunculate oaks (*Quercus robur*), at most three to four metres tall, are rooted between great granite boulders. Wistman's was also put forward by Frank Oliver who, in addition, proposed Tandridge Hill, Surrey.[47]

Although several more lists were compiled by various bodies in the intervening years, many of the same locations appeared on both the SPNR and Huxley lists; in some cases an area, rather than a particular site within that area, being specified in the latter. Some differences between the lists arose because changes had happened to the sites in the intervening years, usually as a result of human activity. Other differences were attributable to altered objectives, in particular, resulting from the slowly dawning recognition that conservation of wildlife was needed, not mere preservation. Also, in the intervening years zoologists had come to accept that there was no need to set up reserves to protect single species; if vegetation was protected then, automatically, the habitat of a rare species of bird, butterfly, or moth, would be preserved. (Little attention was paid during the same years to the possibility that some plant species might be endangered.) When Wicken Fen first came under the control of the National Trust there was a general idea that it should be allowed to run wild and, thereby, to return to its original state. Attitudes had completely reversed by the time the BES published 'Nature conservation and nature reserves' in 1944, with the recognition that 'a great deal of the vegetation which it is desired to preserve is partly the result of human activity through the centuries'.[48] There was a new realisation that, 'If particular species associated with a cultural landscape were to be conserved, it was essential to manage them by continuing or reintroducing traditional management'.[49]

It might have been expected that Tansley would oversee the Wild Life Committee's work from a distance, adopting the role of the wise and elderly statesman, standing back from practical aspects of selecting potential nature reserves. Far from it. As many sites as possible should be seen by members of the committee at first hand, he decided, and he, Tansley, wanted to be part of that exercise – besides, he loved being out in the field with his fellow biologists! With Watt and Fitter, he went back to the Breckland (where they discovered part of the dune system at Lakenheath Warren had been destroyed by the construction of an airfield). With Fitter and others, he assessed potential reserves along the Suffolk coast, rejecting Thorpeness because it was too near unfenced roads and was used by the public for boating, but approving Minsmere because of its outstanding birdlife (Tansley was particularly impressed by its brackish water habitats, which cried out for ecological studies).[50]

On establishment of the Nature Conservancy, the plan was that it would either own reserves (after purchase or gift), lease them, or enter into a Nature Reserve Agreement whereby the landowner would manage the site according to guidelines laid down by the Conservancy. With Tansley as Chairman until January 1953, and Diver as Director General[51] (until December 1952, when he was succeeded by the more energetic and politically astute Max Nicholson), the Conservancy made steady progress in acquiring land, despite the meagre and disappointing budget allowed by successive post-war governments whose priorities lay elsewhere. By 1963, 47 National Nature Reserves had been declared, and by 1975, 99 had been established in England and 41 in Scotland, comfortably exceeding the expectations of the Huxley and Ritchie Committees.[52]

Progress in setting up research centres was much more difficult but, in large part thanks to Nicholson's skilful manoeuvrings, the Conservancy's first research station was opened in 1954. It was at Merlewood, in Cumbria, carefully selected for its position close to two areas of exceptional interest, the north shore of the mudflats of Morecambe Bay and the southern woods and waters of the Lake District. Shortly afterwards, in 1954, a second station was opened at Furzebrook, near Wareham, where the ecology of both the Dorset heaths and the 'Jurassic Coast' could be studied with equal ease.

Tansley could see in the Nature Conservancy the fulfillment of one of his long-cherished aims, a body representing his science which, in all but name, functioned as a research council. There would be henceforward an organisation that could select those areas of Britain most in need of conservation, and that could commission research to find out how conservation was best achieved.

Bringing ecology to a new audience

Even while progress towards the establishment of a Nature Conservancy was finally being made, Tansley was not content to rest for, typically, he saw in the broader picture a challenge. Recognising the growth of public interest in the countryside and wild places of Britain, which had begun in the 1930s and accelerated in the immediate post-war years, and realising that the practice of ecology would need increasing numbers of trained men and women, Tansley devoted much of his time in retirement to introducing ecology to a wider, and more importantly, a younger audience.

His ambition was not exactly new, for as long ago as 1923, when he wrote *Practical Plant Ecology: A Guide for Beginners in Field Study of Plant Communities*,[53] he had sought to make the subject understandable and attractive to young men and women. Reserved by nature and according to some not the most exciting lecturer, in retirement he used his considerable talents as a writer to promote the subject, leaving the more personal task of face-to-face recruitment to younger and more charismatic men. In his Oxford years he had given a great deal of help to the Science Masters' Association.[54] Now, in 1946, he revised *Practical Plant Ecology* (1923), renaming it *Introduction to Plant Ecology*. Also in the same year, he published *Plant Ecology and the School*, written in collaboration with a schoolteacher and close friend, E. Price Evans. Tansley involved a practising teacher not merely to ensure the book was pitched at the right level but also because he greatly admired the example set by Price Evans, who based all biological teaching in his small school in North Wales on practical studies of a nearby woodland.

Tansley and Price Evans followed their introductory chapters with 'brief descriptions of the main types of vegetation that can be studied by teacher and pupils and of what can be done in them'[55], emphasising all the while that 'ecology is essentially a *practical* study'.[56] However, acknowledging that in many schools in towns or cities the students could not go to the plants, they cited the example of the ingenious Lilian Clarke who had created a teaching resource in the grounds of her South London school (Chapter 9), and argued that ecology could be still be taught with profit if plants were brought to the students.[57] An enthusiastic book reviewer in the *Schoolmaster* magazine commented on *Plant Ecology and the School*, 'I cannot imagine that any teacher or student can peruse its pages without feeling a desire to go out into Nature's domain, and learn more of her wonderful ways'.[58] Clearly, the authors' message was hitting the right targets.

Writing *Britain's Green Mantle* (1949) was a much longer project, taking Tansley several years to complete. Similar in overall structure and content to *The British Islands and their Vegetation*, whose primary audience was professional ecologists and knowledgeable amateurs, *Green Mantle* was lighter in weight and tone, avoiding technical terms and Latin names as far as possible. Its audience was the post-war public, or rather that fraction of it possessing open, enquiring minds and a basic love of nature. Its simple aim was to popularise ecology. Tansley's method was to help readers to understand 'our native vegetation'[59] and thereby enhance their enjoyment of it. If, along the way, some were encouraged to take an active part in conservation then so much the better. *Green Mantle* was in its simplicity and easy appeal similar in style to a new series of books beginning to appear at about that time under the collective title, 'New Naturalist' (see p. 174).

* * *

The village of Burwell borders Wicken Fen, and lies less than ten miles northeast of Grantchester and Cambridge. During the 1930s and early 1940s, Burwell's general practitioner was Dr Eric Ennion, a keen ornithologist and highly talented artist, well known among the local community of naturalists. In this same period, Francis Butler was an inspector of sciences in schools of the London County Council. However, when London's children were evacuated to the countryside to escape the wartime blitz, Butler was given responsibility for the education of children billeted in the Cambridge–Newmarket area. One deficiency that soon struck Butler was that his evacuees, like the local children, were never taken out into the Cambridgeshire countryside to study the rich variety of plants and animals on their doorstep. Butler recognised the school curriculum was crowded but concluded that the main obstacle to field studies was a lack of confidence among the teachers who themselves knew precious little about the natural world. Butler's vociferous protests led to him being introduced to Ennion who, in turn, introduced him to a small group of academic biologists in Cambridge, including Tansley and Godwin. This group was already discussing the establishment, as soon as the war was over, of well-equipped residential centres where schoolteachers could be shown how to teach field studies.

A proposal soon emerged to form a Council for the Promotion of Field Studies and an inaugural meeting was held in the British Museum (Natural History) on 10 September 1943. Among those in attendance were Clapham, Salisbury, and Watt. The minutes record that Tansley argued for the inclusion of geography and geology in any programme of development since those subjects underpinned all aspects of field biology. Tansley was appointed to the newly formed Executive Committee and Butler, with his knowledge of the educational establishment, was appointed its Secretary.

Seeking support from the SPNR, Butler wrote next day to its Secretary, Herbert Smith, explaining:

> Children are keen on studying living plants and animals in their natural environment and it is coming to be realised among educationists that this aspect of the subject needs to be encouraged, but, unfortunately, few teachers – including even university graduates in the biological sciences – have the requisite first-hand knowledge and experience of field-work to teach Natural History with competence and enthusiasm.

The solution, explained Butler, was:

> to create a certain number of Hostels for Field Studies, in appropriate localities, each under a trained Warden. ... such a Hostel [would be] able to provide adequate board and lodging for some 30 to 40 students [with] a common room, a field-laboratory, a field museum and a library.
>
> I have put my scheme before the National Trust and they are very sympathetic to the proposal.

It was Butler's hope that 'some, if not all, the Hostels might be established on National Trust property'.[60]

Butler's hopes were to a large extent realised. Events moved rapidly. At the first formal meeting of the Executive Committee, held in Clare College, Cambridge, Tansley was elected President of the Council for the Promotion of Field Studies, and the zoologist, Charles M. Yonge was elected Chairman, though when he moved from a university chair in Bristol to one in Glasgow in 1945 he was replaced by W. H. Pearsall. The Field Studies Council, as it was soon renamed, was able to lease properties from the National Trust. Its first centre, at Flatford Mill on the Essex–Suffolk border, opened in 1946. Godwin was the chairman of Flatford Mill's management committee[61] and the Centre's warden – the man who had persuaded the National Trust to buy the mill – was the warm, outgoing, Eric Ennion. He had gladly given up his medical practice and agreed to work for the first year unpaid because he was so enthused by the project (luckily, he had a small income from writing and broadcasting). Centres at Juniper Hall (Surrey) and Malham Tarn (Yorkshire) soon followed, both leased from the National Trust. Dale Fort (Pembrokeshire) was leased on favourable terms from its owners, the West Wales Field Society. As Tansley and Price Evans had recommended in *Plant Ecology and the School*, the centres were 'ideal places at which open-air ecological work can be carried out by the upper forms of schools', especially those schools from large towns far away from suitable vegetation.[62] As exemplified by Juniper Hall, an 18th century house close to Box Hill National Nature Reserve and the North Downs Way, the buildings were often of historic interest and centres were usually located in areas of exceptional natural beauty, both factors enriching visitors' experience.

Grants to help meet the costs of modifying and equipping the buildings were obtained from charitable bodies such as the Carnegie Trust and the Goldsmith's Company. The Ministry of Education supported the costs of the Field Studies Council's headquarters in London, as well as the stipends of the wardens. For many students in the 1940s and 1950s, it was their first experience of living away from home; they had to pay £4.20p (equivalent to £100 today) a week towards the costs of their board and lodging and the wages of the domestic staff. This ended in 1950 when the Ministry decided, as in the end ministries always do, that the centres should be self-financing, so student fees had to rise.[63]

The drive and personality of the wardens was the key to the success of the whole field studies movement. They were a disparate group but each was an energetic enthusiast, a role model for their students. John Barrett (warden at Dale Fort), who read history at university and was converted to field studies only when he shared a prisoner of war camp with a group of keen ornithologists, exemplified their outlook.[64] 'We [wardens] would try to show ... how different parts of the natural world hung together – plants, animals, rocks, geography, the climate. All of us ... understood exactly why G. M. Trevelyan had said that a historian needed a stout pair of boots as much as a library'.[65]

C. C. Fagg, the first warden at Juniper Hall, was a retired Customs and Excise Officer. It can be surmised that Tansley was instrumental in Fagg's selection for the two had known each other since 1913 when, as a founder member of the BES, Fagg had helped to draft its rules dealing with admitting local societies to corporate membership. He had been since 1906 a vigorous member of the Croydon Natural History and Scientific Society, a society to which Tansley had twice given guest lectures on practical aspects of field work. Interested in medical psychology from a very early age, his interest was fuelled by Tansley's *The New Psychology*, and by letters that he, Fagg, had exchanged with Sigmund Freud. When, on Freud's advice, Tansley began the analysis of a patient in Cambridge, but found he did not have the time to complete the task, Fagg took over the patient with success, he claimed.[66] Courses at Fagg's Juniper Hall included not only geology but also sociology, reflecting his active interest in that subject and his belief in its relevance to ecology.

To support the fledgling Field Studies Council, Tansley again took up his pen, writing for the Council a pamphlet, *What is Ecology?* (1951),[67] and a small book, *Oaks and Oak Woods* (1952),[68] beautifully illustrated by Eric Ennion. The pamphlet did more than explain what ecology was, for, by ranging over contemporary problems from the atom bomb to over-population, it demonstrated the value of applied ecology.

> Just as a fundamental knowledge of physics and chemistry is necessary for engineering and for modern industrial processes, and of human anatomy and physiology for surgery and medicine, so a knowledge of outdoor biology, soil science and climatology, which may best be integrated and expressed as ecology, is necessary for any industry concerned with growing plants, such as agriculture and forestry – for all use of land which contemplates maintenance of the existing natural vegetation or its replacement by some crop of man's choosing.[69]

By 2010, the Field Studies Council had 14 residential centres and three centres for day-courses scattered throughout the United Kingdom. Although, on becoming its Chairman, Pearsall found the organisation 'hopelessly involved' and had 'the unpleasant task of curtailing expansion and reducing administrative costs and aims to practical levels',[70] the organisation survived. Its website proudly records today that, since its foundation, over 2.5 million people have attended, learned from, and enjoyed its courses.

The Field Studies Council, like the Nature Conservancy, and the national parks, was both a product and a cause of an upsurge in interest in the natural world that had started in the 1930s and gained momentum in the 1940s. Now, on the eve of the 1950s, this interest was fed by broadcasters, such as Ennion, and by publications, such as *Britain's Green Mantle* and the hugely popular 'New Naturalist' series, whose authors taught without condescension and respected their readers. Topics for the latter series were selected by an editorial board that included distinguished biologists, such as Julian Huxley and John

Gilmour, and professional communicators, such as James Fisher. The first volume to appear was E. B. Ford's *Butterflies*, in 1945, while other early contributors to the series were Max Nicholson, W. H. Pearsall, Richard Fitter, and E. J. Salisbury.

Recognition

In the New Year's Honours list published by *The Times* on 2 January 1950 there appears under the heading Knights Bachelor, 'Tansley Professor Arthur George FRS, Chairman of Nature Conservancy'. His family believe he accepted the knighthood only reluctantly and in response to the urging of his daughters, who forcefully reminded him that the title 'Lady Tansley' would help Edith's work as a county councillor, which she loved so dearly. Perhaps, too, his acceptance was Edith's reward for her unswerving loyalty and support.

Two other honours seem to have given Tansley himself just as much pleasure. The first was his election in 1928 to the Athenaeum under the little used Rule 2 which provided for the admission of men 'of distinguished eminence in Science, Literature, Arts or for Public Service'.[71] The second came in 1944 when, with the support of G. M. Trevelyan, the Master, he was made an Honorary Fellow of Trinity, his old Cambridge college. He was able, henceforward, to dine in Trinity whenever he chose. Trevelyan also helped paved the way to a long overdue rapprochement between Tansley and Bertrand Russell. The two elderly men were able to heal the wound that had separated them for too many years of their lives, Russell admitting about their long ago spat, 'you were right then, Tansley, and I was wrong'.[72]

Age and increasing frailty finally caught up with Tansley when in 1953, aged 83, he reluctantly conceded that he had to reduce his commitments. Increasing deafness finally forced him to relinquish the Chairmanship of the Nature Conservancy and also of the Field Studies Council.[73] He had largely fulfilled his ambitions. His services to his nation, and to posterity, had been recognised. Only two years later he would be dead.

Notes

1. Godwin 1939, p.549.
2. Tansley 1939b, p.527.
3. Sheail 1976, p.86.
4. Ibid., p.160.

5. The Association for the Preservation of Rural Scotland was founded in 1927, and the Council for the Preservation of Rural Wales in 1928 (Sheail 1998, p.13).
6. Sheail 1976, p.71: Sheail 1998, p.13.
7. Sheail 1976, p.72.
8. H. C. Cowles campaigned for more than a decade before a small part of the Indiana Dunes was designated a State Park in 1926, protecting it from encroachment by steel mills and power plants. Not until 1966 was the much larger Indiana Dunes National Lakeshore created (Cassidy 2007, p.68).
9. Robertson 2003, p.00.
10. Brown 2005, p.104.
11. After reading agronomy at Cornell University, Leonard Elmhirst worked as an agricultural economist, spending many years in India.
12. Sheail 1998, p.26.
13. Overy 2009, p.80.
14. Olby 2004, p.93.
15. Zuckerman 1987, p.163.
16. Sheail 1976, p.91; Sheail 1998, p.10.
17. Sheail 1976, p.95; Sheail 1998, p17.
18. Sheail 1998, p.18.
19. Sheail 1976, p.100; Sheail 1998, p.186.
20. Anon. 1944. Nature conservation and nature reserves. *Journal of Ecology* **32**, p.45–82.
21. Ibid., p.82.
22. Tansley AG. 1943. Nature reserves. *The Spectator*, 519–520.
23. Trevelyan 1929, p.22.
24. Tansley 1945, preface.
25. Ibid., p.6.
26. Ibid., p.41.
27. Pauline Dower replaced her husband, John, when he became too ill to serve on the Committee. She had been closely involved in his work; as his health deteriorated, she had acted as his chauffeur on site visits.
28. Huxley 1970, p.288.
29. Nominations to the Hobhouse and Huxley Committees were initiated by Silkin, on advice from civil servants and expert advisers, such as Huxley. Huxley and Lt Colonel E. N. Buxton were the only two who sat on both Committees (contrary to what Huxley implies about Tansley).
30. Sheail 1976, p.112.
31. Nicholson 1987, p.93.
32. Sheail 1998, p.30.
33. The Act did not extend to Scotland, except with respect to nature conservation (Sheail 1976, p.205).
34. Sheail 1976, p.215.
35. Blackmore 1974, p.427.
36. Sheail 1976, p.224.

37. Sheail 1976, p.204; Sheail 1998, p.16, p.31, p.34.

38. An exclusive gentleman's club, located in London's Pall Mall. Populated mainly by men of inherited wealth, status, or, in Prime Minister Arthur Balfour's words, of 'undiluted distinction' (Darwin 1943, pp.27–28).

39. The quotation comes from Richard Fitter's draft, kindly shown to me by his son, Alastair. In the final published version (Fitter 1989, p.203) reference to the Athenaeum was omitted.

40. Sir Charles Trevelyan was an MP and member of MacDonald's Labour Cabinets of 1924 and 1929–1931 (Cannadine 1992, p.11). See also note 36, Chapter 4.

41. An ecological geneticist, Ford was an infamous misogynist who respected few women. An exception was Miriam Rothschild, a distinguished zoologist and granddaughter of Charles Rothschild, founder of the SPNR.

42. Tansley 1945, p.57; John B. Whittaker, personal communication.

43. Archives of the University of Oxford, Bodleian Library, URB/BG/1 file 1.

44. Huxley 1973, p.241.

45. Crowfoot 1991, pp.95–96.

46. Tansley 1949, p.93.

47. Rothschild, Marren 1997, pp.38–39. Staffhurst Wood was made a Local Nature Reserve by Surrey County Council in 1972; Wistman's Wood was made a Forest Nature Reserve in 1956, and later a National Nature Reserve (it is at the heart of Dartmoor National Park); and Tandridge Hill lies on the North Downs Way, a long distance footpath running just south of the M25 motorway.

48. Anon. 1944. Nature conservation and nature reserves. *Journal of Ecology* **32**, 45–82.

49. Briggs 2010, p.365.

50. Sheail 1976, pp.151–152.

51. Under Diver, the Conservancy lacked clear direction – it even failed to produce a statutory annual report – causing Tansley to take a more 'hands-on' role than he would have wished (John Sheail, personal communication).

52. Sheail 1976, p.216.

53. Tansley AG. 1946. *Introduction to Plant Ecology*. London: Allen & Unwin (a revision of *Practical Plant Ecology*, 1923).

54. Godwin 1958, p.4.

55. Tansley, Price Evans 1946, p.9.

56. Ibid., p.7.

57. Ibid., p.31.

58. Tansely 1949, dustjacket.

59. Tansley 1949, p.v.

60. Butler 1943, pp.3–4.

61. Godwin 1985b, p.201.

62. Tansley, Price Evans 1946, p.8.

63. Tansley AG, Butler FHC. 1950. Funds for field studies. *The Times* 17 October.

64. Berry 1988, p.2; Crothers 2000, p.550.

65. Barrett 1987, p.37.

66. Cameron, Forrester 2000, pp.235–237.

67. Tansley AG. 1951. *What is Ecology?* Council for the Promotion of Field Studies (reprinted 1987 in *Biological Journal of the Linnean Society* **32**, 5–16).

68. Tansley AG. 1952. *Oaks and Oak Woods*. London: Methuen.

69. Tansley 1951 (see note 66), p.12.

70. Pearsall manuscript, Kendall Records Office, Cumbria, 7 February 1955.

71. Godwin 1977, p.17; *The Times* 1 May 1928, p.19.

72. Godwin 1977, p.19.

73. Godwin 1957, p.240.

12 A Detached Liberal Philosopher and Free-thinker

Harry and Margaret Godwin were among Tansley's oldest and dearest friends. Tansley had helped to shape not only Harry's career but that too of Margaret, of whom he was very fond. In about 1930, he brought to her attention a new technique for pollen analysis recently developed in Sweden, pointing out how quantification of the various tree pollens in the different strata of peat sediments could reveal the history of post ice age (Quaternary) vegetation. Through the analysis of soil cores, or cylinders, collected by boring into peat, such as that at Wicken Fen, it would thus be possible to find not only which species dominated at any particular time within a span of thousands of years, but when and by what other species those dominants were succeeded. To the list of examples of plant successions which Tansley was accumulating might be added cases involving considerably longer timescales – all thanks to the remarkable decay resistance of pollen grains. Margaret immediately followed Tansley's advice and before long made herself an internationally respected authority on vegetational history.[1]

The closeness between the young Godwins and Tansley shines out one of Harry's anecdotes:

> I well recall a dinner in [London's] Soho called by A.G. [Tansley] one evening in the spring of 1929, to enable the Godwins to decide in what part of Europe they might take their holiday beginning on the following morning. Unobtrusively

Shaping Ecology: The Life of Arthur Tansley, First Edition. Peter Ayres.
© 2012 by John Wiley & Sons, Ltd. Published 2012 by John Wiley & Sons, Ltd.

excellent waiting at the Cordon Bleu Italian restaurant made us innocents familiar with the facility with which Chianti removes control of the nether limbs. It
was only with mutual support that we reached the Haymarket ...[2]

Not surprisingly, it was Godwin who was invited both by the British Ecological
Society (BES)[3] and by the Royal Society[4] to write obituaries of Tansley shortly
after his death in 1955. And it was Godwin, too, who was invited in 1975 to
deliver the first of the BES's prestigious biennial Tansley Lectures, appropriately
choosing to talk about, 'Sir Arthur Tansley: the man and his subject'. Chronicling
the major events and achievements in Tansley's life with a clarity for which this
and any other biographer must be grateful, Godwin's collected judgements are
those of a close friend and self-confessed admirer of his subject. In fairness, he
is occasionally critical, recording that Tansley was occasionally harsh with
others, proved difficult with his children, and failed to appreciate the freedom
he enjoyed by virtue of his inherited wealth.[5] In the main, however, he eulogises
Tansley, avoiding any critical views of his professional life or hint of the pain he
inflicted on his family.

This final chapter attempts to reach a more balanced and fairer view of an
undeniably remarkable man. It includes the views of those who were more critical
than Godwin of Tansley's contribution to the rise of ecology and conservation,
and reminds us of those limitations to Tansley's character and work seen earlier.

* * *

Failure was almost unknown in Tansley's life. A rare exception was the project
that sought to produce a New Students' British Flora (NSBF). In allowing himself to become involved with it, Tansley showed a rare lapse of judgement, even
though in the end he did manage to salvage something of worth, albeit a very
different Flora.

The need and potential market for a new Flora was recognised in February
1933 by Oxford University's Clarendon Press. The book which the new Flora
would replace was the *Manual of British Botany*, written by Charles Babington
in1843 and in its 10th edition after revision by A. J. Wilmott in 1922.[6] The new
Flora would be comprehensive but compact, thoroughly up to date, and would
be edited from draft contributions submitted by a group of 30 experts (in the
event, 44), each invited to write about their 'own' group of plants. The Director
of Kew, Sir Arthur Hill, was supportive, suggesting that his deputy, John
Gilmour, should be one of the editors. Hill acknowledged that the British
Museum should also be involved, even though his own institution and the
Museum were not always happy bedfellows, and suggested the obvious person
to represent the Museum would be its deputy keeper of botany, A. J. Wilmott.
Probably looking for a 'neutral' and respected referee with political skills, Hill
suggested that Tansley should be asked to join and chair what would become

thereby an editorial committee of three. Although he was at the time heavily committed to leading botany in Oxford, to researching and writing *The British Islands and their Vegetation*, and by his own admission was no taxonomist, Tansley undertook the chairmanship of the project on the understanding that he would not be involved in its day to day routines. Progress was initially rapid, the committee holding its first meeting on 26 May 1933.[7]

The reasons why the huge and cumbersome project failed have been set out fully by Briggs and Gorringe (2002) but, in brief, they include delays in sending invitations to contributors, coupled with a lack of clarity in the guidance given to contributors by the editors, particularly where such notoriously difficult taxonomic groups as *Hieracium* (hawkweeds) and *Rosa* (roses and briars) were concerned. There were disagreements between editors, ranging from the fundamental, such as which classification scheme was to be used – Tansley first opposed Gilmour's recommended scheme and then changed his mind – to the trivial but niggling issue of the appropriate remuneration for contributors. To make matters worse, in the summer of 1933 Tansley suffered a rare period of ill health following the removal of most of his teeth. The editors' timetable started to slip. Soon Gilmour and, more so, Wilmott, who had been allocated the greatest number of plant groups to deal with, were pleading that overwork was delaying their contributions to the project. A proposal to enlarge the committee to five and thereby accelerate progress was opposed by Tansley, although he was himself either unable or unwilling to give to the project the time it needed. In the summer of 1935 he even ducked the issue of Wilmott's tardiness, instead urging the diplomatic Gilmour to chivvy Wilmott. A contrast of personalities lay at the heart of the project's problems for, whereas Gilmour was kind hearted and courteous, Wilmott 'was volcanic, unpredictable',[8] a ferocious perfectionist. He had already been involved in a failed collaborative venture, the Cambridge British Flora project, conceived by his ex-supervisor, the forthright and sometimes difficult C. E. Moss. Wilmott had found the bulk of the work falling on his shoulders when Moss emigrated to South Africa.[9] With this unhappy precedent, and not by inclination a team player, Wilmott harboured plans to write his own flora, an unrealised ambition which must have sapped his enthusiasm as he worked on the NSBF.

Whatever life remained in the project, the outbreak of World War II (WWII) in 1939 crushed it as many contributors were recruited elsewhere to help fight the war. Tansley wrote, 'I do not see that there need or should be any idea of *abandoning* the work'[10] but he was probably merely trying to lighten spirits during increasingly dark days. No correspondence concerning the NSBF has been found that is dated later than September 1945, when Tansley was suggesting to Gilmour that the Flora should be published in parts, as each was completed. By this date, however, Tansley had conceived an alternative plan, whereby something of value could be rescued from the NSBF's wreckage, a smaller flora designed to meet the needs of ecologists.

Humphrey Gilbert-Carter, Director of the University Botanic Garden in Cambridge, was a regular visitor at Grove Cottage in Grantchester. On Sunday afternoons he would walk across the meadows from the city to take tea with the Tansleys. One Sunday, probably in the winter of 1944–1945, he was walking in the meadows with Tom Tutin (Professor of Botany at Leicester University) when he suggested, in a way that appeared impromptu but which was probably carefully planned, that they should see if Tansley was at home. He was, and he invited them to stay to tea.

Tansley urged Tutin to get together with Roy Clapham and Edmund 'Heff' Warburg (Oxford) to write a flora for the use of students. It would not be for the expert, so would not compete with the 'big flora' (the NSBF), but would be 'a book suitable for students in Universities to enable them to identify the plants they meet on excursions with reasonable ease and accuracy.'[11] Both Tutin and Warburg had been on the list of contributors to the NSBF and they, like several of the botanists who were to help them with the new project, were quick to see the virtue in Tansley's plan. It would capitalise on the effort they had already put into the NSBF, drawing on material which in all likelihood would otherwise be wasted. The outcome was Clapham, Tutin, and Warburg's hugely successful *Flora of the British Isles* (1952)[12], and the shorter, pocket-sized *Excursion Flora of the British Isles* it spawned in 1959.[13]

Maybe Tansley should not have strayed into the world of taxonomy, for whose politics he had little feeling. Or, maybe, he simply had too much work on his shoulders in the mid-1930s. Whatever the case, for once in his life he was not totally committed. His vacillation over the choice of the taxonomic basis for the NSBF occurred at about the same time he was vacillating with regard to plans to move the Department of Botany in Oxford, suggesting that even his exceptional mind and energies could occasionally be overtaxed.

* * *

In his Inaugural Lecture at Oxford in1927,[14] Tansley had spoken not just about the future of botany within the university but of the relevance of his own subject to the wider world. On a more philosophical note, he addressed fundamental questions about the place of science in the modern world, the financial dependence of scientists upon government, and the consequent responsibilities of scientists towards government. He advocated, unequivocally, freedom for scientists. An 'essential condition of successful development is freedom from utilitarian *compulsion* … good effective research cannot be done to order, nor can the spirit of investigation be manufactured artificially. That spirit depends absolutely on the driving force of natural curiosity'. Tansley's credo was to be seriously threatened in the next decades.

Alongside discussions of the need for more rigorous economic planning in the 1930s, there was a growing body of eminent scientists who argued that science should be driven not by curiosity but by prevailing social and economic conditions. Their views were expressed most eloquently in J. D. Bernal's *The Social Function of Science* (1939),[15] in which he argued that Soviet-style planning was needed to harness science for the greater social good.[16] WWII, which harnessed large numbers of scientists to tasks ranging from the development of radar to the employment of computers in code-breaking, merely served to emphasise the interdependence of the state and its scientists. The Oxford zoologist, John Baker, was so worried by this trend that in 1940 he wrote to 49 British scientists, including Tansley, to express his concern. In Tansley he found a natural ally and, together with Michael Polanyi, Professor of Physical Chemistry in Manchester (a Jewish émigré from Hitler's Germany), they founded the Society for Freedom in Science in order to combat the direction of science towards the state's priorities.[17] Tansley helped draft the Society's statements of belief, among which was the recognition that teamwork (something that was, as yet, unfamiliar to ecologists), or organised activity, need not compromise freedom. Tansley was the Society's first acting Chairman and later was its Vice President. He was active in its affairs until a few weeks before his death in 1955. Writing a brief obituary of Tansley for the Society's *Pamphlet*, Baker said of him and of the period:

> No longer young in years, his quiet enthusiasm rose above every obstacle. The chief difficulty was to make the the new Society's views known. So prejudiced were the times ... that it was found almost impossible to get anything published if it involved opposition to the reigning totalitarian outlook. To start a new journal was illegal under war-time regulations, and even the distribution of typed circulars was made difficult by the restrictions on the sale of paper.

> In those days it was an inspiration to work with Professor Tansley. He was always a good counsellor, urging restraint when wild schemes would have damaged our cause, energetically supporting strong action whenever it could be successful ... When the war came to an end and we could expound our principles freely by the publication of our Occasional Pamphlets, his interest did not lessen.[18]

The copy of the obituary held in the library of the University of Cambridge has been annotated by Harry Godwin. His handwritten notes suggest hidden ambitions among the 'Bernalists' which to modern eyes are astonishing, as was their near acceptance by the Royal Society.

> I think the crucial battle was fought at the big meeting in the [Royal Society Scientific Information Conference, 1948] at which the 'Bernalists' sought to get

general approval for a plan to take-over all reprints produced by British scientific journals, and have a central cttee (totalitarian) distribute them to <u>all</u> research workers needing them. (The owners of journals meeting costs of production). The other officers of the Royal Socy were sold on the idea & Salisbury was on the platform.

The Royal Society Scientific Information Conference had been called to address what were generally agreed to be the current problems of scientific publishing – its slow speed, readers' difficulties in identifying relevant papers because of incomplete abstracting and indexing, and then their difficulties in acquiring the journal or a reprint of a paper of interest once it had been identified. (Readers' difficulties were even greater if they were working outside Europe or North America.) As the founder of two journals, one of which, *New Phytologist*, operated as a non-profit-making charity, and the other of which, *Journal of Ecology*, generated modest but desperately needed surpluses that supported the wider activities of the BES, no subject could have been closer to Tansley's experiences and concerns.[19]

In a pre-emptive move, designed to gain public sympathy for their views, and to forestall the 'Bernalists', Baker and Tansley had written a letter to *The Times* making the case that scientific publication must be free from the stifling effects of rigid central control. 'The journal … will cease to exist, at least in its present form',[20] they claimed, if the totalitarian proposals of the Bernalists were accepted. They cleverly implied that Bernalism threatened not just journals but any learned society whose journal provided a significant part of its income. Their letter was published on Monday 21 June 1948, the very day of the conference, and on the same page *The Times* carried a leading article that could only have delighted Baker and Tansley. It 'anticipated' that the proposals for centralisation and regimentation 'will evoke strong and justified protest', and concluded the conference had to address itself to such 'cavalier and insidious suggestions'. Bernal withdrew his paper, which he had circulated before the conference; the Society for Freedom in Science and the learned societies were victorious.

A vehicle for publicising the Society's views had presented itself to Tansley when he was invited by the University of Oxford to deliver the 1942 Herbert Spencer Lecture. To receive such an invitation was an honour for anyone, but for Tansley it held peculiar significance because Spencer's *The Factors of Organic Evolution* (1887) was one of the first books he had owned and cherished. Moreover, when Tansley was at the outset of his academic career – only 24 years old and a lowly Assistant Professor at University College, London (UCL) – he had been asked by Spencer to help him revise his best-selling *Principles of Biology*, first published in 1864. (Spencer had initially approached Frank Oliver but Oliver had turned him down, at the same time recommending his young

protégé, Arthur Tansley.[21]) Working with the 75-year-old celebrity[22] had proved
a chastening experience, as Spencer repeatedly had to reign in the young man's
enthusiasm:

> I find now, however, that the thinking over biological questions in anticipation is
> very much interfering with my work at present in hand and tends, by giving me
> additional subjects of thought, to derange my health still more.
>
> *H. Spencer to A. G. Tansley, 24 March 1896*

> I wish to make my own scheme [of plant classification] and then to have you aid
> in the revision of it; and to carry out this idea it is better for me that I should not
> see what you have [already] sent.
>
> *H. Spencer to A. G. Tansley, 30 November 1896*[23]

The archives do not reveal the final extent of Tansley's input to Spencer's published revision but in 1942 he gladly delivered the Memorial Lecture, 'The values of science to humanity'. While pointing out the contributions that biology
had made 'to the two great practical arts of medicine and agriculture' and that
we were now 'on the verge of far-reaching social developments in which the
applications of biological and sociological knowledge will play a commanding
part',[24] and also acknowledging the 'scientific importance of the enormous
amount of work that is now done in technological laboratories attached to
industrial firms, instituted or assisted by Government, and sometimes attached
to universities', his lecture was a broad restatement of the ideals of the Society
for Freedom in Science. Curiosity was, he said, 'a primitive human instinct';[25]
'the highest type of research, that which has been productive of the most fundamental discoveries, is essentially the work of individual minds, freely dealing
with their own chosen material'.[26]

* * *

> … a prime condition for satisfactory adjustment of rival claims is that each side
> should really understand the needs of the other and be ready to concede their validity.
>
> *A. G. Tansley to* The Times, *3 December 1946*[27]

It is impossible to quarrel with Godwin's summation of Tansley as 'The epitome
of the detached liberal philosopher and free-thinker'.[28] Tansley displayed those
qualities in activities ranging from his practical involvement in the Magdalen
Philosophy Club, where he joined the realists arguing in favour of a physical
basis to psychology and the latter subject's relation to philosophy, to his defence
of the freedom of the individual scientist. His mind knew no boundaries.
Unravelling the intricacies of one subject simply advanced the understanding
of others. Notably, understanding how the normal, healthy human mind
becomes disturbed can, he believed, help us understand how normal succession

in vegetation can be deflected. In opposing the South African holists' idea of a 'life force in matter' – a concept that appealed to the Oxford idealists – he drew upon his unrivalled knowledge of ecology, dismissing the superorganism and putting in its place the 'ecosystem', our conception of which has changed little since his day.

As mentioned earlier, Tansley often seemed detached in the company of his peers; typically quiet but ready to change the course of a conversation by the interjection of a few carefully chosen words, the product of his reflections. Naturally lacking all ostentation, he was consequently not an exciting public speaker – unless the listener enjoyed the intellectual challenge of analysing his arguments – but he was a prodigious and gifted writer. Words were important to him. Carefully chosen, they formed the framework upon which he constructed ecology. Thus, in addition to 'ecosystem', he was responsible for coining such stalwarts of ecology as 'autogenic' and 'allogenic' succession,[29] 'deflected succession', and 'plagiosere'.[30] In 2007 Stephen Trudgill analysed Tansley's 'The use and abuse of vegetational concepts and terms' (1935) for the series 'Classics in Physical Geography Re-visited', concluding 'Tansley had a scholarly approach which it is difficult to gainsay'.[31]

The number of research papers he published was, even by the standards of his day, unexceptional. His contributions to original research on the effects of management practices on Wicken Fen, and of the effects of grazing on the chalk grasslands of the Downs, were significant, but his primary influence stems rather from his many articles and books that display his unequalled ability to review information, recognise patterns, and point the way forward. Above all, he taught plant ecologists to ask, 'What forces are driving the vegetational changes that we see everywhere we look?'

A generation of young botanists had reason to be grateful for his editorial skills, for his 'unselfish and benificent advice'.[32] More than half the authors of original papers in some early editions of both *New Phytologist* and the *Journal Ecology* were from either UCL or Cambridge University, or were in some other way linked personally to Tansley. There has never been any suggestion that he suppressed views at odds with his own, but his editorial skills were frequently employed in clarifying evidence or augmenting arguments that took ecology in the direction he favoured.

A case in point is the accumulation of evidence in favour of what he called 'anthropogenic climaxes'. As he pointed out in *The British Islands and their Vegetation* (1939a),

> their recognition has been one of the chief advances in the study of British vegetation during the twenty eight years that have elapsed since *Types of British Vegetation* was written. ... it was largely the work of Farrow (1916, 1917) on the

influence of rabbits of the vegetation of Breckland that called general attention to its significance. Watt's work (1919, 1923) on the influence of mice, birds and other small animals in the destruction of tree seeds and seedlings and the consequent prevention of oak and beech regeneration in our woods was also an important contribution in the same direction.[33]

He might have added that his own studies of rabbit grazing in the Ditcham Park estate, begun in 1909 and extended with Adamson to include a swathe of the South Downs, were equally important.

What Tansley's modesty similarly forbade him to mention was his own part in Farrow's and in Watt's studies. In a footnote to the first of E. Pickworth Farrow's series of eight papers, 'On the ecology of the vegetation of Breckland',[34] published in the *Journal of Ecology* between 1915 and 1919, Farrow acknowledged, 'The writer is deeply indebted to Mr. A.G. Tansley for much constant encouragement and frequent advice during the progress of this work and for very kindly reading the manuscript and to suggesting many improvements therein'.[35] (He also thanked, among others, Mr R. S. Adamson, for suggesting the work was needed, and Dr C. E. Moss and Professor F. W. Oliver for their encouragement and inspiration.) In similar vein, Watt, at the conclusion of his *Journal of Ecology* paper, 'On the causes of failure of natural regeneration in British oakwoods' (1919), added 'to Mr Tansley, who suggested the subject to me and who helped me by kindly criticism, advice and discussion, I am especially indebted'.[36] Watt's association with Tansley had begun in 1914. After a first degree taken at the University of Aberdeen, Watt was awarded a Carnegie Scholarship which he was preparing to take up in Germany. The outbreak of war prevented this so, instead, he went to Cambridge to study under Tansley. The outcome of those first researches was not published until 1919 because of Watt's absences from Cambridge for various reasons, including war service during which he was badly gassed in France. It was only after his early discharge from the army that Watt was able to return to Cambridge and complete his studies.

If the support that Tansley gave to Farrow and Watt resulted in him being able to pursue through others his own agenda for ecology, and their successes enhanced in small ways his own reputation, the question has to be asked, 'Was Tansley a cold-hearted schemer, guided by self-interest?' The answer must be an emphatic 'No.' The help he gave to Woodruffe-Peacock (Chapter 9), the interest he took in Pearsall's career (Chapter 10), and the help he gave to Farrow and Watt, might all be judged by a cynic to be examples of self-serving, since they advanced 'his' subject, ecology, but the support and friendship he gave to Zuckerman (Chapter 9) cannot be seen as anything other than generous and kind. Zuckerman was not an ecologist, so his success or failure could have had no effect on Tansley's subject.

In the late 1930s growing numbers of Jewish scientists escaped Hitler's darkening shadow. After Leo Brauner had been dismissed by the University of Jena, Tansley offered him a post in Oxford. Brauner was a plant physiologist, something desperately needed by Tansley, who even arranged a research grant for him. However, within three months, and before the grant arrived, Brauner accepted the offer of a chair at the University of Istanbul. In spite of his feelings of disappointment, Tansley generously gave £100, most probably from his own pocket, to help the departing Brauner with his travel and maintenance costs.[37] Hans Heller was a neuroendocrinologist who left the University of Vienna for a safer but uncertain future in Britain. For a while Heller and his family found shelter with the Tansleys at Maycoes, their holiday home in Branscombe (M. Tomlinson, personal communication). Eventually, Heller obtained a lectureship at the University of Bristol where, in 1949, he was made Professor of Pharmacology.

Similar kindness, with no reward expected, was displayed towards the family of Thomas Chipp, Kew's Deputy Director. When Chipp died in June 1931 at the young age of 44, leaving a wife and two children, Tansley swiftly assumed a leading role in coordinating efforts to raise money to help the family. He wrote to Sir Arthur Hill, 'Chipp was so widely appreciated and admired, and there were so many people who had a warm affection for him that I feel sure you will get a generous response.'[38] An appeal was circulated later in 1931 among Chipp's friends and colleagues. In spite of it being printed on paper headed 'Royal Gardens, Kew', Hill's position precluded him from being a signatory. However, there was no shortage of sponsors. In addition to Tansley, they included A. C. Seward, F. E. Weiss, Lt-Colonel Sir David Prain, an ex-Director of Kew, and a number of other ex-soldiers connected with Chipp's earlier military service. Tansley generously promised to contribute £30 (equivalent to over £1400) each year, again from his own pocket. By the end of 1931, £654 had already been raised to help Mrs Chipp and to provide for the education of her children.[39]

All the evidence is that to his fellow biologists, if not to his family, Tansley was kind and supportive, always ready to dispense wise advice and inspiration. Where ecology and young ecologists were concerned, he was, paradoxically, even paternalistic.

* * *

Few of Tansley's contemporaries had anything but praise for him. Dissent from the general opinion, where it occurred, is often traceable to envy. So it was with Isaac Bayley Balfour, who loathed Tansley. His negative feelings stemmed not just from the 'Manifesto' affair but, as described in Chapter 8, the way Tansley had positioned himself at the head of ecology. And so it was also with

E. J. Salisbury who on the occasion of celebrations marking the 50th anniversary of the British Ecological Society blatantly downplayed Tansley's role in its formation and growth. In 'The origin and early years of the British Ecological Society' (1964), Sir Edward (a founder member of the BES) quite reasonably traced the origin of vegetation studies in Britain to the brothers William and Robert Smith, adding to the names of these pioneers two others, Patrick Geddes and D'Arcy Thompson.[40] 'In actual fact', continued Salisbury, 'it was William Smith who, in 1904, called together a meeting to consider the formation of a 'Central Committee for the Study and Survey of British Vegetation'.[41] The founder members are not listed by Salisbury and the only reference to Tansley is as the editor of *Types of British Vegetation* (1911b), 'in preparation for the International Phytogeographic Excursion'. Even if the truth about Tansley's part in the formation of the Central Committee were to lie closer to Salisbury's than to Godwin's account, Tansley's leading role in the initiation, organisation, and leadership of the Excursion, which is ignored by Salisbury, is well attested by overseas visitors such as Clements and Cowles, both of whom, unlike either Salisbury or Godwin, were witnesses to the event.

After praising Oliver, who 'gave breadth to British ecology which the French school [as under Flahault] for some time lacked',[42] Salisbury recounted, 'it was under Oliver, to whom he became an assistant, that Tansley became interested in this aspect of plant study and he brought to the subject an editorial gift and an organising ability that was invaluable and often enabled him *to reap where others had sown*' [present author's italics]. Salisbury's treatment of Tansley was a perfect example of the cliché, 'to damn with faint praise'. While acknowledging that Tansley's experience with the *New Phytologist* had been helpful when the BES was planning its *Journal of Ecology*, Salisbury writes, 'actual editing was initially undertaken by Dr Cavers but his tenure of this office was cut short by illness and early death from cancer. Tansley thereupon generously filled the gap'.[43] Thus, Salisbury attributes Tansley's assumption of the editorship to chance rather than a well thought out plan, in doing so overlooking the contrary evidence outlined here in Chapter 7. Given Salisbury's eminence and his distinguished record as a botanist, his coldness towards Tansley, verging on negativity, deserves exploration.

Fourteen years younger than Tansley, Salisbury's life had many parallels with that of the older man. Salisbury had from 1905 been a student at UCL where he was taught by Oliver, and he had been a regular member of the summer excursions to both the Bouche d'Erquy and, from 1910 onwards, to Blakeney.[44] He won a Quain studentship and two of his earliest papers were published with Oliver, one on Palaeozoic seeds, the other on the ecology of *Suaeda fructicosa* (shrubby sea-blite) on shingle beaches.[45] Other early papers by Salisbury were published by Tansley in his *New Phytologist*. It was while Salisbury was a research student that he was invited to join the British Vegetation Committee. As his research matured he took a special interest in coastal vegetation and woodland ecology, in 1921 publishing jointly with Tansley on the subject of oak woodlands near Malvern. Salisbury's work was characterised by his emphasis on making measurements of the physical environment, especially light, and by his quantitative approach, especially to seed production and viability. Succeeding Oliver in 1929 as the Quain Professor of Botany at UCL, he took the Directorship of Kew in 1943. In spite of his hugely successful career, there was scope for Salisbury to feel frustrated, to wonder whether his career might have been even more successful had it not been for Tansley, who in so many instances had got there first!

Like Tansley, Salisbury had had a long involvement in the struggle to establish national nature reserves. He was on the drafting committee of the 'Conference on Nature Preservation in Post-war Reconstruction' convened by the Society for the Promotion of Nature Reserves (SPNR), which in its submission to government in 1941 recommended the establishment of nature reserves (Chapter 11). He was also a member of the Nature Reserves Investigation Committee (NRIC) that was subsequently set up to provide evidence to government, providing his own carefully constructed list of existing and proposed reserves.[46] He was a member of the BES's committee, under Tansley's chairmanship, that identified sites for nature reserves; indeed there is a good case that Salisbury coordinated the pressure from within the Royal Society (RS) which helped get that committee established. Further, as Biological Secretary of the RS from 1945 to 1955, he would have been regularly involved in meetings between representatives of government and the learned societies; in particular, he was responsible for presenting the RS's submission to the Huxley Committee.[47] Salisbury and Tansley both had dealings in another sphere, for the former was President (1938–1944) of the School Nature Union, which helped to set up the Field Studies Council. It must have seemed to Salisbury that always it was Tansley who held the senior position, always it was Tansley who received the plaudits. Maybe it was a case of his general irritation boiling over when, responding to the letter that Tansley and Baker had sent to *The Times* about the Royal Society's conference on science publishing, Salisbury let it be widely known that he 'deplored that an essentially private issue should have been dragged into public dispute'.[48]

Both men spent much time sitting on committees, but there their behaviour was very different. Whereas Tansley's interventions were few and pithy, Salisbury 'delighted in talking and often launched into monologues, always interesting and frequently witty and amusing, though liable to be deemed overlengthy by the impatient listener'.[49] Such differences need not, however, be a cause for dislike. In writing his obituary of Salisbury, Clapham notes, 'He was normally friendly and helpful … but could at times seem unsympathetic or even overbearing', a view which harmonises with the memories of Clapham's student, Donald Pigott, who recalls Salisbury 'had difficulties' not just with Tansley but with a number of other leading ecologists, including Diver and Elton (C. D. Pigott, personal communication). Taking inordinate pride in his Directorship of Kew, Salisbury had an intense dislike of Oxbridge, its influence, and all those, probably including Tansley, who were associated with the old universities. Given such feelings, it is little wonder that in 1964 the recently retired Salisbury gave minimal praise to Tansley's role in the early years of the BES.

Whereas Tansley and Druce treated each other with respect, albeit without a great deal of admiration (Chapter 9), Tansley's relationship with another Oxford colleague was probably more difficult. Arthur Church was already in his sixties and had been overlooked for the Sherardian Chair in Oxford which Tansley won in 1927. Church had been a Fellow of the Royal Society since 1921 on the strength of morphological research which he illustrated, brilliantly, himself. A notoriously awkward character, he had had many set-backs in his life, both professional and personal, which probably sharpened his biting wit. One Oxford student recalled that, 'at an appropriate moment in lectures, Tansley became the butt of Church's jokes. "Tansley and Miss [Edith] Chick – *pause* – looked down the same microscope – *pause* – Miss Chick is now Mrs. Tansley"'.[50] It was an example of what Church liked to call his Principle of Propinquity: 'Put a couple in the same laboratory and propinquity does the rest'. When late in his career he turned his attentions to ecology, Church was ignored by the ecological establishment (not just by Tansley), in large part because, as Tansley put it, 'he felt that observations unrelated to large generalising schemes of thought, in which he delighted, were of little value or significance'.[51] No mention is made in *The British Islands* of Church's ecology, betraying Tansley's view of its value. After retiring in 1930 – which spared Tansley too many years of direct contact – Church died in 1937. As one of the most knowledgeable but least travelled botanists of his generation, he was, in Tansley's words, 'a genius manqué'.[52]

Comparing the two men, Mabberley (2000) concluded that Church was much the greater botanist. This was true in the narrow sense, of being able to recognise plants, of being familiar with his local flora – Church was the author of *Introduction to the Plant-life of the Oxford District* – and of having an exhaustive knowledge of floral anatomy. And Tansley never hid the fact that he himself had great difficulty in remembering and recognising plant species in the field.[53] But, in the broader sense, of their influence on the international development of botany, Tansley made a contribution that far exceeded that of Church and, indeed, of almost every other man or woman of his generation.

* * *

Tansley's other self-recognised weakness was that he was no experimental-ist, though as seen in Chapters 7 and 8 he may have been a little too hard on himself. A weakness, if there was one, may have derived from shortcomings in his education where the planning and execution of experiments was concerned, or it may have stemmed from the deformity of his left hand which restricted his manual dexterity. Whatever the causes of Tansley's deficiencies, his response was to develop strengths in other areas. Eschewing the sociological approach to vegetation which was dominating ecology in continental Europe, and which required highly developed taxonomic and floristic skills, often linked to an inti-mate knowledge of the local flora, he took British ecology in another direc-tion.[54] Recognising the dynamic nature of vegetation, his ecology was devoted to understanding the causes of change, whether they stemmed from the natural environment or, as he found in so many situations, from human activity. In spite of his own weakness as an experimentalist, Tansley recommended to others an experiment-based, quantitative approach to ecological problems; an approach that sought explanation through integrated environmental and plant physiological measurements.

Ever the man with the overview, liberal in his outlook, and alert to the vir-tues of other people's ways of looking at the world, Tansley never dismissed the value of the phytosociology that grew from the Zürich and Montpellier schools of, respectively, Schröter and Flahault. In a 1922 review for the *Journal of Ecology*,[55] he attempted to help British ecologists understand the European viewpoint by translating from French or German into English some of the more common terms used by the sociologists. He emphasised how much the British and continental botanists had to learn from each other. In 1925 he contributed an article to Schröter's *Festschrift*,[56] explaining how an under-standing of vegetation on the chalk soils of southern England was only enhanced when the two approaches, the 'dynamic' British and the 'classifica-tory' Continental, were brought together. He was at this time publishing, with Adamson, their studies of the chalk Downs. Their 1926 paper contains a lengthy section (pp.24–31) with the heading 'Sociological constitution of chalk grassland', in which data presented earlier in the paper are reanalysed according to the precepts of the phytosociologists. As letters received from Clements in the early 1930s show,[57] he and Tansley regretted the gulf between the Anglo-American and Continental traditions. They speculated whether a small, strictly informal, meeting between leaders from the two traditions – Braun-Blanquet and Du Rietz were regarded as the key phytosociologists[58] – might be able to resolve major misunderstandings and differences. Sadly, for reasons unknown, their speculations remained just that because the meeting was never arranged. However, in 1939, Tansley, Clapham, and Watt were asked by the BES to select two continental sociologists to attend the Society's next

Annual Meeting. That attempt at bridge-building was thwarted by the outbreak of WWII.[59]

A set of events after the war display not only Tansley's respect for phytosociologists but also his liberal mind set. They relate to Reinhold Tüxen, who in Zürich in 1926 had met and been inspired by Braun-Blanquet, and who in the years immediately preceding WWII become the leading phytosociologist in Germany.[60] Tüxen had in many people's eyes become associated with the Nazi party. This was not because of any political sympathy or affiliation – indeed, he helped a number of French prisoners of war escape – but because he had given advice on plantings alongside the *Reichsautobahn* (motorway network) being constructed in the 1930s. This advice probably won him the support he needed to develop a prestigious research institute in Hanover (Zentrallstelle für Vegetationskartierung des Reichs (Centre for Vegetation Mapping for the Empire)) of which he was Director.[61] After the war, Tüxen concluded that Tansley had been directly responsible for persuading the British occupying powers to grant him repossession of the institute. He thanked Tansley in person when they met at the International Phytogeographic Excursion in Ireland in 1949. Through the ecological advances his team made, particularly in the 1950s, Tansley's support for Tüxen and the Centre was more than repaid.[62]

Shortly before his death, Tansley was still lamenting the lack of harmony between the British and Continental approaches.[63] His contributions to ecology are, however, now widely appreciated in Europe, as well as in North America. Christian Körner, one of Europe's leading plant ecologists, and the holder of the Schimper Chair in Basel since 1989, wrote for the Tansley website,[64] 'He changed what was formerly descriptive biogeography, or plant sociology, into what – with the help of later workers such as MacArthur, Harper, Grime, and others – would become the new Anglo-American functional ecology'. He had shaped modern ecology.

Contributing to the same website, Michael Huston (Texas State University), a world authority on biodiversity, wrote:

> Ecology has developed a different 'flavor' on different continents, at least partially because of differences in environmental conditions. The density-independent, disturbance-driven concepts of Andrewartha and Birch, which were developed in the unproductive and drought-prone ecosystems that cover much of Australia, were often seen as conflicting with the more deterministic, competition-driven ideas of American ecologists who worked in the productive grasslands and forests of North America. British ecology, developed on a landscape without the great extremes of conditions seen in Australia or North America, was more nuanced, focusing on the details and mechanisms underlying more subtle variations in vegetation.

By mid-century, each continental school of ecology had its Founders and its Giants. Arthur Tansley was certainly the first Giant of British Ecology, and ranks among the global Giants of Ecology in terms of his lasting impact on the field around the world.

Whereas, in the years following World War I, the leadership of some botanical disciplines, such as physiology, were passing to the USA,[65] Tansley's endeavours kept Britain at the forefront of the world of ecology where, happily, thanks to the impetus he bequeathed, it remains today.

On balance, the evidence suggests Godwin has handed down to us a fair assessment of his friend and mentor. Tansley *was* detached, both from colleagues, who regretted he was hard to know but respected the soundness of his judgement, and from his family, for whom he was 'a taker, an absorber, a receiver' (M. Tomlinson, personal communication). He was definitely a philosopher, both by nature and by application. And if the essence of being a liberal free-thinker is the abhorrence of coercion and regimentation, he was that too. Tansley's thinking defined a new scientific discipline but, in the institutions he left behind, his legacy was also a practical one – something that can be said about too few thinkers.

* * *

In the final months of his life Tansley was still busy with correspondence, typically dispensing help and advice to fellow ecologists. In the spring of 1955 he was discussing with A. S. Thomas of the Nature Conservancy the rival claims of the yews at Kingley Vale and at Great Yews, near Salisbury, to be the oldest and broadest in England. In late summer he was sending to Max Nicholson his comments on the draft of that year's Nature Conservancy Annual Report.[66] At the Bristol meeting of the British Association for the Advancement of Science (BAAS), held in September, he gave the introductory talk at a symposium organised around a theme close to his heart, chalk grassland. It was, fittingly, his last major talk. Only two months later he was admitted to Addenbrooke's Hospital in Cambridge suffering from prostate cancer. A major operation to remove the cancer was planned but never happened because his condition worsened following a preliminary operation to relieve its symptoms.[67] Arthur Tansley died at his home in Grantchester on 25 November 1955. He was 84 years old. Edith survived him by 15 years, managing to reach the age of 100, and active almost to the end.

How would he want to be remembered? He described himself as ' a dilettante with a strong interest in science and philosophy, some considerable analytical and critical talent, and good powers of exposition'.[68] The idea of a biography would almost certainly have embarrassed this very private man but, perhaps, if that biography ends among 'the great hills of the South country' his spirit might be mollified. In Kingley Vale's well-ordered patterns of shrub and grassland, its broad sweeping views, and its remote loneliness, it is possible to sense echoes of Arthur Tansley. It was from Kingley Vale that he drew early inspiration and it was to the protection of Kingley Vale and its like that the long road of his life took him back.

Notes

1. Godwin 1985b, p.155.
2. Godwin 1977, p.25.
3. Godwin 1958.
4. Godwin 1957.
5. Ibid., p.241.
6. Briggs, Gorringe 2002, p.2.
7. Ibid., p.3.
8. Ibid., p.11.
9. Allen 2010, p.371.
10. Briggs, Gorringe 2002, p.9.
11. Ibid., p.10.
12. Clapham AR, Tutin TG, Warburg EF. 1952. *Flora of the British Isles*. Cambridge: Cambridge University Press.
13. Clapham AR, Tutin TG, Warburg EF. 1959. *Excursion Flora of the British Isles*. Cambridge: Cambridge University Press.
14. Tansley 1927.
15. Bernal JD. 1939. *The Social Function of Science*. London: George Routledge.
16. Pioneer X-ray crystallographer, Desmond Bernal FRS ('Sage' to his contemporaries) was educated in an Irish Jesuit seminary. Godwin (1985b, p.194) suggested that the adult Bernal substituted the convictions of Marxism for those of Catholicism.
17. Godwin 1985b, pp.192–3; Bocking 1997, pp.26–7.
18. Baker JR. 1955. *Sir Arthur Tansley 1971–1955*. Occasional Pamphlet No. 16. London: Society for Freedom in Science.
19. A consortium of learned societies submitted a memorandum to the Royal Society stating, 'the proposed scheme of central publication and distribution … would interfere with the statutory and accepted aims of individual societies as free centres of interest and encouragement of research' (Brown 2005, p.293).
20. Baker, Tansley 1948, p.5.
21. Tansley Archives, University of Cambridge Library.
22. Spencer was at the height of his popularity in the late 1870s and 1880s, when his book sales approached 1 million. By the 1890s, his views on social Darwinism – the use of evolutionary arguments, such as his term 'survival of the fittest', to justify economic and social inequities – had begun to tarnish his reputation.

23. Tansley Archives, University of Cambridge Library.
24. Tansley 1942, p.107.
25. Ibid., p.105.
26. Ibid., p.110.
27. 'Training areas', a letter from Tansley to *The Times*, 3 December 1946, concerning the rival claims on the countryside made by the military and conservationists.
28. Godwin 1977, p.24.
29. Tansley 1929b, p.680.
30. Godwin 1977, p.22.
31. Trudgill 2007, p.519.
32. Godwin 1958, p.3.
33. Tansley 1939a, p.129. Here, Tansley undervalued Watt's contribution; the focus on regeneration was novel and significant. Watt (1947) showed regeneration was fundamental to cycles within plant communities – cycles that in communities such as beechwoods included a critical gap phase where mature trees were absent from sizeable patches.
34. While a student of botany, Farrow had been introduced to psychology by Tansley, who later helped him make professional connections. In the 1930s and 1940s, Farrow authored several research papers and a popular book designed to help readers with self-analysis (Cameron and Forrester, 2000). Breckland is a gorse-covered, sandy heath of 940 square kilometres, lying northeast of Cambridge but south of the Fens. It is one of the driest places in Britain.
35. Farrow 1915, p.211.
36. Watt 1919, p.203.
37. Morrell 1997, p.374.
38. Archives of the Royal Botanic Garden, Kew, TF Chipp letters, ff91, Tansley to Mead.
39. Ibid., ff97, November 1931.
40. Geddes and Thompson were among the first professors at University College, Dundee, but were odd choices, unless it is because they tutored the Smith brothers (see note 41, Chapter 6). Although Patrick Geddes was a well-respected plant geographer, and was in 1880 perhaps the first to use the word 'ecology' in its contemporary (interdisciplinary) sense, he was a renaissance man, his career encompassing many subjects, not least sociology. D'Arcy Wentworth Thompson was a distinguished, mathematically inclined, biologist not known for studies of vegetation.
41. Salisbury 1964, p.13.
42. Ibid., p.14.
43. Ibid., p.17.
44. On joining his first Bouch d'Erquy party, Salisbury was not yet an advanced student. His precocious talents had been noticed by Oliver who offered him a special invitation (Clapham 1980, p.505).
45. Clapham 1980, p.537.
46. Sheail 1976, p.142.
47. Sheail 1998, p.27.
48. Tansley Archives, University of Cambridge Library; copy of Baker 1955 (see note 18) annotated by H. Godwin.

49. Clapham 1980, p.537.
50. Mabberley 2000, p.117.
51. Tansley 1939c, p.434.
52. Ibid., p.443.
53. Godwin 1977, p.23.
54. Allen 2010, p.372.
55. Tansley AG. 1922. The new Zurich–Montpellier school. *Journal of Ecology* **10**, 241–248.
56. Tansley AG. 1925. The vegetation of the southern English chalk (Obere Kreide-Formation). Festschrift Carl Schröter. *Verüffentlichungen des Geobotanischen Institutes Rübel in Zürich* **3**, 406–430.
57. H53 Tansley Archives, University of Cambridge Library.
58. Josias Braun-Blanquet studied with both Schröter and Flahault before becoming unofficial leader of the Zürich–Montpellier School of Phytosociology. In analysing vegetation he put emphasis on pattern rather than process but, as he confided to an eminent colleague, Heinz Ellenberg, even he was at times frustrated by the effort his followers put into the minutiae of classification, at the expense of thinking about dynamics and processes (P. J. Grubb, personal communication). The Swede, Einar Du Rietz, independently established a School of Phytosociology at Uppsala (the Nordic School).
59. Sheail 1987, p.162.
60. Barkman 1981, p.88.
61. Tüxen's help was enlisted by the landscape architect Alwin Seifert who, in tune with Nazi ideology, advocated the planting of native species, stating, 'We declare picea pungens glauca [Colorado blue spruce] enemy of the state No.1'. *Pinus montana* [mountain pine] was 'a sin against the nobility of our mountains' (Zeller 2007, pp.153–154).
62. Sheail 1987, p.162.
63. Tansley 1954, p.viii.
64. www.newphytologist.org/tansley/
65. Ayres 2008, p.188.
66. H24 Tansley Archives, University of Cambridge Library.
67. Tansley to his daughter, Margaret Tomlinson, 2 November 1955, Branscoll (a collection of family letters found at Branscombe, South Devon).
68. Godwin 1957, p.241.

References

Adamson RS. 1912. An ecological study of a Cambridgeshire woodland. *Journal of the Linnean Society, Botany* **40**, 339–384.

Adamson RS. 1918. On the relationships of some associations of the Southern Pennines. *Journal of Ecology* **6**, 97–109.

Allen DE. 1986. *The Botanists*. London: St Paul's Bibliographies.

Allen DE. 1994. Oxford and Druce. *Botanical Society of the British Isles News* **67**, 41–45.

Allen DE. 2010. *Books and Naturalists*. London: Collins.

Anker PJ. 1999. *The ecology of nations*. PhD thesis, University of Harvard.

Anker PJ. 2001. *Imperial Ecology: Environmental Order in the British Empire, 1895–1945*. Cambridge, MA: Harvard University Press.

Anker PJ. 2002. The context of ecosystem theory. *Ecosystems* **5**, 611–613.

Armstrong P. 2000. *The English Parson-Naturalist*. Leominster: Gracewing.

Ayres PG. 2005. *Harry Marshall Ward and the Fungal Thread of Death*. St Paul, MN: American Phytopathological Society.

Ayres PG. 2008. *The Aliveness of Plants. The Darwins at the Dawn of Plant Science*. London: Pickering & Chatto.

Baker JR, Tansley AG. 1948. Threat to journals. Resolution at to-day's conference. *The Times* 21 June, p.5.

Barkman J. 1981. Reinhold Tüxen 1899–1980. *Vegetatio* **48**, 87–91.

Barrett JH. 1987. The Field Studies Council: how it all began. *Biological Journal of the Linnean Society* **32**, 31–41.

Barton R. 1990. 'An influential set of chaps': the X-club and Royal Society Politics 1864–1885. *British Journal for the History of Science* **23**, 53–81.

Bastian H. 2008. Lucy Wills (1888–1964). The life and research of an adventurous woman. *Journal of the Royal College of Physicians, Edinburgh* **38**, 89–91.

Berry RJ. 1988. Natural history in the twenty-first century. *Archives of Natural History* **15**, 1–14.

Blackman FF, Oliver FW, Blackman VH, Keeble F, Tansley AG. 1917. The reconstruction of elementary botanical teaching. *New Phytologist* **16**, 241–252.

Blackman FF, Tansley AG. 1905. Ecology in its physiological and topographical aspects. *New Phytologist* **4**, 199–203, 232–253.

Blackmore M. 1974. The Nature Conservancy: its history and role. In: *Conservation in Practice*. Eds A Warren, FB Goldsmith. Chichester: John Wiley, pp.423–436.

Bocking S. 1997. *Ecologists and Environmental Politics*. New Haven, NY: Yale University Press.

Boney AD. 1991. The 'Tansley manifesto' of 1917 – *plus ça change* …? *New Phytologist* **118**, 1–21.

Boney D. 1995. The botanical 'Establishment' closes ranks: fifteen days in January 1921. *The Linnean* **11**, 26–37.

Bower FO. 1918. Botanical bolshevism. *New Phytologist* **17**, 105–107.

Bower FO. 1938. *Sixty Years of Botany in Britain (1875 –1935). Impressions of an Eye-witness*. London: Macmillan.

Briggs D. 2010. *Plant Microevolution and Conservation in Human-Influenced Ecosystems*. Cambridge: Cambridge University Press.

Briggs D. Gorringe E. 2002. The struggle to produce a Flora of the British Isles (1933–1952). *Watsonia* **24**, 1–15.

Briggs GE. 1948. Frederick Frost Blackman. 1866–1947. *Obituary Notices of Fellows of the Royal Society* **5**, 651–658.

Brockliss LWB. 2008. *Magdalen College, Oxford. A History*. Oxford: Magdalen College.

Brown A. 2005. *J. D. Bernal. The Sage of Science*. Oxford: Oxford University Press.

Butler FHC. 1943. Copy of the letter which started it all. *Biological Journal of the Linnean Society* **32**, 3–4.

Cameron L. 1999. Histories of disturbance. *Radical History Review* **74**, 4–24.

Cameron L, Forrester J. 1999. A nice type of the English scientist: Tansley and Freud. *History Workshop Journal* **48**, 64–100.

Cameron L, Forrester J. 2000. Tansley's psychoanalytic network: an episode out of the early history of psychoanalysis in England. *Psychoanalysis and History* **2**, 189–256.

Cameron L, Matless D. 2011. Translocal ecologies: The Norfolk Broads, the "Natural" and the International Phytogeographic Excursion, 1911. *Journal of the History of Biology* **44**, 15–41.

Cannadine D. 1992. *G. M. Trevelyan. A Life in History*. London: Harper Collins.

Cassidy VM. 2007. *Henry Chandler Cowles. Pioneer Ecologist*. Chicago: Sigel Press.

Chick E, Tansley AG. 1903. On the structure of *Schizaea malacanna*. *Annals of Botany* **17**, 493–510.

Clapham AR. 1980. Edward James Salisbury. 16 April 1886–10 November 1978. *Biographical Memoirs of Fellows of the Royal Society* **26**, 502–541.

Clements ES. 1960. *Adventures in Ecology*. New York: Pageant Press.

Clements FE. 1905. *Research Methods in Ecology*. Lincoln, NB: The University Publishing Company.

Clements FE. 1912. The International Phytogeographic Excursion in the British Isles. VIII. Some impressions and reflections. *New Phytologist* **11**, 177–179.

Clements FE. 1916. *Plant Succession*. Washington, DC: Carnegie Institute.

Cowles HC. 1901. The physiographic ecology of Chicago and vicinity: a study of the classification of plant societies. *Botanical Gazette* **31**, 73–108.

Cowles HC. 1912. The International Phytogeographical Excursion in the British Isles. IV. Impressions of foreign members of the party. *New Phytologist* **11**, 25–26.

Cowles HC. 1919. Current literature: book reviews. *Botanical Gazette* **68**, 477–478.

Creese MRS. 1998. *Ladies in the Laboratory? American and British Women in Science, 1800–1900*. London: Scarecrow Press.

Crothers J. 2000. JH Barrett MBE, MA, Msc, 1913–1999. *Field Studies* **9**, 549–554.

Crowcroft P. 1991. *Elton's Ecologists. A History of the Bureau of Animal Population*. Chicago: University of Chicago Press.

Darley G. 1990. *Octavia Hill. A Life*. London: Constable.

Darwin B. 1943. *British Clubs*. London: Collins.

Davies, L. 1904. *The Working Men's College 1854–1904. Records of its History and its Work for Fifty Years by Members of the College*. London: Macmillan.

Doyle AC. 1912. The lost world. *Strand Magazine* (published serially, April to November 1912). (Reprinted 2008 in Oxford World's Classics. Oxford: Oxford University Press.)

Druce GC. 1886. *Flora of Oxfordshire*. Oxford: Parker & Co.

Endersby J. 2008. *Imperial Nature. Joseph Hooker and the Practices of Victorian Science*. London: University of Chicago Press.

Falcon-Lang H. 2008. Marie Stopes: passionate about paleaobotany. *Geology Today* **24**, 132–136.

Farrow EP. 1915. On the ecology of the vegetation of Breckland: I. General description of Breckland and its vegetation. *Journal of Ecology* **3**, 211–228.

Fischedick KS, Shinn T. 1993. The International Phytogeographic Excursions, 1911–1923: intellectual convergence in vegetation science. In: *Denationalising Science*. Eds ET Crawford, S Sörlin. Kluwer: Dordrecht, pp.107–131.

Fitter RSR. 1989. Fifty years of wildlife conservation in Britain – a personal view. *Oryx* **23**, 202–207.

Fitzgerald P. 1984. *Charlotte Mew and Her Friends*. London: Collins. (Republished 1992 by Flamingo Books).

Foster JB. 2000. *Marx's Ecology*. New York: Monthly Review Press.

Godwin H. 1939. Report on the Easter meeting 1939. *Journal of Ecology* **27**, 549–550.

Godwin H. 1957. Arthur George Tansley 1871–1955. *Biographical Memoirs of Fellows of the Royal Society* **3**, 227–246.

Godwin H. 1958. Sir Arthur George Tansley FRS, 1871–1955. *Journal of Ecology* **46**, 1–8.

Godwin H. 1977. Sir Arthur Tansley: the man and his subject. *Journal of Ecology* **65**, 1–26.

Godwin H. 1985a. Early development of *The New Phytologist*. *New Phytologist* **100**, 1–4.

Godwin H. 1985b. *Cambridge and Clare*. Cambridge: Cambridge University Press.

Golley, F. 1993. *A History of the Ecosystem Concept*. New Haven, CT: Yale University Press.

Green JR. 1914. *A History of Botany in the United Kingdom from the Earliest Times to the End of the 19th Century*. London: J. M. Dent.

Griffin N. 1992. *The Selected Letters of Bertrand Russell 1884–1914. The Private Years.* London: Houghton Mifflin.

Grugeon A. 1873. *Botany: Structural and Physiological.* London: Murby.

Gunther RT. 1904. *A History of the Daubeny Laboratory Magdalen College, Oxford.* Oxford: Henry Frowde.

Hale Bellott H. 1929. *University College London. 1826–1926.* London: University of London Press.

Harrison B. 1994. *The History of the University of Oxford. VIII. The Twentieth Century.* Oxford: Clarendon Press.

Harrison JFC. 1954. *A History of the Working Men's College 1854–1954.* London: Routledge & Kegan Paul.

Harrison JFC. 1961. *Living and Learning 1790–1960. A Study in the History of the English Adult Education Movement.* London: Routledge & Kegan Paul.

Hill TG. 1909. The Bouche d'Erquy in 1908. *New Phytologist* **8**, 189–195.

Hoskins WG. 1955. *The Making of the English Landscape.* London: Hodder & Stoughton.

Hutchinson J. 1946. *A Botanist in South Africa.* London: P. R. Gawthorn.

Huxley J. 1970. *Memories*, vol. I. London: Allen & Unwin.

Huxley J. 1973. *Memories*, vol. II. London: Allen & Unwin.

Hywel-Davies J, Thom V. 1984. *The Macmillan Guide to Britain's Nature Reserves.* London: MacMillan. (The authors cite Max Nicholson in *Report of the Nature Conservancy* for the year ended 30 September 1958, p.91.)

Kelly T. 1992. *A History of Adult Education in Great Britain.* Liverpool: Liverpool University Press.

Langenheim JH. 1996. Early history and progress of women ecologists. *Annual Review of Ecology and Systematics* **27**, 1–53.

Le Mire ED. 1969. *The Unpublished Lectures of William Morris.* Detroit: Wayne State University Press.

Lester J. 1995. *E. Ray Lankester and the Making of Modern British Biology.* London: British Association for the History of Science.

Lévêque C. 2003. *Ecology from Ecosystem to Biosphere.* Enfield, NH: Science Publishers.

Lewis CS. 1991. *All My Road Before Me: The Diary of CS Lewis 1922–27.* London: Harper Collins.

Mabberley D. 2000. *Arthur Harry Church. The Anatomy of Flowers.* London: Merrell and Natural History Museum.

MacCarthy F. 1994. *William Morris. A Life of Our Time.* London: Faber & Faber.

Marilaun K von. 1896. *The Natural History of Plants. Their Forms, Growth, Reproduction and Distribution* (English translation and revision by F. W. Oliver). London: Blackie.

McIntosh RP. 1985. *The Background of Ecology.* Cambridge: Cambridge University Press.

Moorman M. 1980. *George Macaulay Trevelyan. A Memoir by His Daughter.* London: Hamish Hamilton.

Morrell J. 1997. *Science at Oxford 1914–1939. Transforming an Arts University.* Oxford: Oxford University Press.

Moss CE. 1910. The fundamental units of vegetation. *New Phytologist* **9**, 18–53.

Moss CE. 1913. *Vegetation of the Peak District*. Cambridge: Cambridge University Press.

Moss CE, Rankin WM, Tansley AG. 1910. The woodlands of England. *New Phytologist* **9**, 113–149.

Mullen R, Nunson J. 2009. *The Smell of the Continent: The British Discover Europe 1814–1914*. London: Macmillan.

Murphy G. 1987. *Founders of the National Trust*. London: Christopher Helm.

Nicholson M. 1987. *The New Environmental Age*. Cambridge: Cambridge University Press.

Olby R. 2004. Huxley, Sir Julian Sorell (1887–1975). *Oxford Dictionary of National Biography* **29**, 92–95.

Oliver FW. 1906. The Bouche d'Erquy in 1906. *New Phytologist* **5**, 189–195.

Oliver FW. 1913. *Makers of British Botany*. Cambridge: Cambridge University Press.

Oliver FW. 1914. Nature reserves. *Journal of Ecology* **2**, 55–56.

Oliver FW. 1927. *An Outline of the History of the Botanical Department of University College, London*. London: University College.

Oliver FW. Tansley AG. 1904. Methods of surveying vegetation on a large scale. *New Phytologist* **3**, 228–237.

Overy R. 2009. *The Morbid Age: Britain Between Wars*. London: Allen Lane.

Pennington J. 1948. *The British Heritage, the People, their Crafts and Achievements as Recorded in their Buildings and on the Face of the Countryside*. London: Odhams.

Phillips JFV. 1931. The biotic community. *Journal of Ecology* **19**, 1–24.

Porter H. 1968. Vernon Herbert Blackman. 1872–1967. *Biographical Memoirs of Fellows of the Royal Society* **14**, 36–60.

Rackham O. 2006. *Woodlands*. London: Harper Collins.

Robertson J. 2003. Max Nicholson. *ECOS. A Review of Conservation* **24**, 52–53.

Rothschild M. 1987. Changing conditions and conservation at Ashton Wold, the birthplace of the SPNR. *Biological Journal of the Linnean Society* **32**, 161–170.

Rothschild M, Marren P. 1997. *Rothschild's Reserves. Time and Fragile Nature*. Colchester: Harley Books.

Russell B. 1946. *A History of Western Philosophy*. London: Allen & Unwin.

Salisbury EJ. 1952. Francis Wall Oliver 1864–1951. *Obituary Notices of Fellows of the Royal Society* **8**, 229–240.

Salisbury EJ. 1964. The origin and early years of the British Ecological Society. *Journal of Ecology* **52** (suppl.), 13–18 (supplement edited by A. Macfadyen and P. J. Newbould).

Salisbury EJ, Tansley AG. 1921. The Durmast oak-woods (*Querceta sessili-florae*) of the Silurian and Malvernian strata near Malvern. *Journal of Ecology* **9**, 19–38.

Sanders D. 2005. From radicle to radical: biology education and the first women Fellows of the Linnean Society of London. *The Linnean* **21**, 20–24.

Schimper AFW. 1898. *Pflanzen-geographie auf Physiologischer Grundlage*. Jena: G. Fischer. (English translation by P. Groom and I. B. Balfour. 1903. *Plant Geography upon a Physiological Basis*. Oxford: Clarendon Press.)

Schulte FK. 2000. From survey to ecology: the role of the British Vegetation Committee, 1904–1913. *Journal of the History of Biology* **33**, 291–314.

Seaward MRD. 2001. E. Adrian Woodruffe-Peacock (1858–1922): a pioneer ecologist. *Archives of Natural History* **28**, 59–69.

Secord A. 1996. Artisan botany. In: *Cultures of Natural History*. Eds N Jardine, JA Secor, EC Spray. Cambridge University Press, Cambridge, pp.378–393.

Sheail J. 1976. *Nature in Trust*. London: Blackie.

Sheail J. 1987. *Seventy-Five Years in Ecology: The British Ecological Society*. Oxford: Blackwell Scientific Publications.

Sheail J. 1998. *Nature Conservation in Britain. The Formative Years*. London: The Stationery Office.

Sheail J. 2002. *An Environmental History of Twentieth-Century Britain*. London: Palgrave.

Smuts JC. 1926. *Holism and Evolution*. London: MacMillan.

Spalding F. 2001. *Gwen Raverat. Friends, Family and Affections*. London: Pimlico.

Spencer H. 1864. *Principles of Biology*, vol. I. (vol. II, 1867). London: Williams & Norgate.

Spencer H. 1887. *The Factors of Organic Evolution*. London: Williams & Norgate.

Steward FC 1947. In memoriam. Frederick Frost Blackman. *Plant Physiology* **22**, ii–viii.

Tanner JR. 1917. *The Historical Register of the University of Cambridge, being a Supplement to the Calendar with a Record of University Honours and Distinctions to the Year 1910*. Cambridge: Cambridge University Press.

Tansley AG. 1896a. The Stelar theory – a history and a criticism. *Science Progress* **5**, 133–150.

Tansley AG. 1896b. The Stelar theory – a history and a criticism. II. The metamorphosis of the stele. *Science Progress* **5**, 215–226.

Tansley AG. 1902. The old and the new "Phytologist" (Editorial). *New Phytologist* **1**, 10.

Tansley AG. 1903. An experiment in ecological surveying. *New Phytologist* **2**, 167–168.

Tansley AG. 1904a. The problems of ecology. *New Phytologist* **3**, 191–200.

Tansley AG. 1904b. A second experiment in ecological surveying. *New Phytologist* **3**, 200–204. (Tansley writing as editor.)

Tansley AG. 1905. Formation of a committee for the survey and study of British vegetation. *New Phytologist* **4**, 23–26. (Tansley writing as editor.)

Tansley AG. 1911a. The International Phytogeographic Excursion in the British Isles. II. Details of the Excursion. *New Phytologist* **10**, 276–291.

Tansley AG. 1911b. *Types of British Vegetation*. Cambridge: Cambridge University Press.

Tansley AG. 1913a. International Phytogcographic Excursion (IPE) in America, 1913. *New Phytologist* **12**, 322–336.

Tansley AG. 1913b. The aims of the new journal. *Journal of Ecology* **1**, 1–3.

Tansley AG. 1914a. International Phytogeographic Excursion (IPE) in America, 1913. *New Phytologist* **13**, 30–31, 83–92, 268–275, 325–333.

Tansley AG. 1914b. Presidential address. *Journal of Ecology* **2**, 194–202.

Tansley AG. 1916. Albert Stanley Marsh. *New Phytologist* **15**, 81–85.

Tansley AG. 1920. *The New Psychology and its Relation to Life*. London: Allen & Unwin.

Tansley AG. 1922a. *Elements of Plant Biology*. London: Allen & Unwin.

Tansley AG. 1922b. Studies of the vegetation of the English chalk: II. Early stages of redevelopment of woody vegetation on chalk grassland. *Journal of Ecology* **10**, 168–177.

Tansley AG. 1923. *Practical Plant Ecology: A Guide for Beginners in Field Study of Plant Communities*. London: Allen & Unwin.

Tansley AG. 1927. *The Future Development and Functions of the Oxford Department of Botany.* Oxford: Clarendon Press.

Tansley AG. 1929a. Obituary notice: William Gardner Smith. *Journal of Ecology* **17**, 172–173.

Tansley AG. 1929b. Succession: the concept and its values. *Proceedings of the International Congress of Plant Sciences, Ithaca* **1**, 677–686. Manasha, WI: Banta.

Tansley AG. 1935. The use and abuse of vegetational concepts and terms. *Ecology* **16**, 284–307.

Tansley AG. 1939a. *The British Islands and their Vegetation.* Cambridge: Cambridge University Press.

Tansley AG. 1939b. British ecology during the past quarter-century: the plant community and the ecosystem. *Journal of Ecology* **27**, 513–530.

Tansley AG. 1939c. Arthur Harry Church. 1865–1937. *Obituary Notices of Fellows of the Royal Society* **2**, 433–443.

Tansley AG. 1942. The values of science to humanity. *Nature* **150**, 104–110.

Tansley AG. 1945. *Our Heritage of Wild Nature. A Plea for Organised Nature Conservation.* Cambridge: Cambridge University Press.

Tansley AG. 1947a. The early history of modern plant ecology in Britain. *Journal of Ecology* **35**, 130–137.

Tansley AG. 1947b. Obituary notice. Frederick Edward Clements, 1874–1945. *Journal of Ecology* **34**, 194–196.

Tansley AG. 1948. The nature and range of variation in the floral symmetry of *Potentilla erecta.* (L.) Hampe. *New Phytologist* **47**, 95–110.

Tansley AG. 1949. *Britain's Green Mantle: Past, Present, and Future.* London: Allen & Unwin.

Tansley AG. 1952. *Mind and Life: An Essay in Simplification.* London: Allen & Unwin.

Tansley AG. 1954. Some reminiscences. *Vegetatio* **5–6**, vii–viii.

Tansley AG, Adamson RS. 1925. Studies of the vegetation of the English chalk: III The chalk grasslands of the Hampshire–Sussex border. *Journal of Ecology* **13**, 177–223.

Tansley AG, Adamson RS. 1926. Studies of the vegetation of the English chalk: IV A preliminary survey of the chalk grasslands of the Sussex Downs. *Journal of Ecology* **14**, 1–32.

Tansley AG, Chick E. 1901. Notes on the conducting tissue-system in Bryophyta. *Annals of Botany* **15**, 1–38.

Tansley AG, Chipp TF. 1926. *Aims and Methods in the Study of Vegetation.* London: Whitefriars Press.

Tansley AG, Fritsch FE. 1905. Sketches of vegetation at home and abroad: the flora of the Ceylon littoral. *New Phytologist* **4**, 1–17, 27–55.

Tansley AG, Price Evans E. 1946. *Plant Ecology and the School.* London: Allen & Unwin.

Taylor D. 1968. *The Godless Students of Gower Street.* London: University College Union.

Tomlinson M. 1983. *Three Generations in the Honiton Lace Trade. A Family History.* Privately published.

Trevelyan GM. 1929. *Must England's Beauty Perish? A Plea on Behalf of the National Trust for Places of Historic Interest or Natural Beauty.* London: Faber & Gwyer.

Trudgill ST. 2007. Classics in physical geography revisited: Tansley A. G. 1935. The use and abuse of vegetational concepts and terms. Ecology 16, 284–307. *Progress in Physical Geography* **31**, 517–522.

Walters SM. 1981. *The Shaping of Cambridge Botany.* Cambridge: Cambridge University Press.

Warming JEB. 1896. *Lehrbuch der ökologischen Pflanzengeographie. Eine Einführung in die Kenntnis der Pflanzenvereine.* Berlin: Borntraeger. (English translation by P. Groom and I. B. Balfour. 1909. *Oecology of Plants: An Introduction to the Study of Plant Communities.* Oxford: Clarendon Press.)

Watt AS. 1919. On the causes of failure of natural regeneration in British oakwoods. *Journal of Ecology* **7**, 173–203.

Watt AS. 1947. Pattern and process in plant communities. *Journal of Ecology* **35**, 1–22.

Willis AJ. 1994. Arthur Roy Clapham, 1904–1990. *Biographical Memoirs of Fellows of the Royal Society* **39**, 73–90.

Willis AJ. 1997. Forum. The ecosystem: an evolving concept viewed historically. *Functional Ecology* **11**, 268–271.

Wilson AN. 2002. *The Victorians.* London: Hutchinson.

Zeller T. 2007. *Driving Germany. The Landscape of the German Autobahn, 1930–1970.* Oxford: Berghahn Books.

Zuckerman S. 1978. *From Apes to Warlords.* London: Hamish Hamilton.

Zuckerman S. 1987. Comments and recollections. In: *Julian Huxley. Biologist and Statesman of Science.* Eds CK Waters, A Van Helden. Houston, TX: Rice University Press, pp. 162–164.

Index

Shaping Ecology: The Life of Arthur Tansley, First Edition. Peter Ayres.
© 2012 by John Wiley & Sons, Ltd. Published 2012 by John Wiley & Sons, Ltd.